WITHDRAWN
HARVARD LIBRARY
WITHDRAWN

Science and Theology in the Reformation

Science and Theology in the Reformation

Studies in Theological Interpretation and Astronomical Observation in Sixteenth-Century Germany

Charlotte Methuen

t&t clark

Published by T&T Clark
A Continuum imprint
The Tower Building, 11 York Road, London SE1 7NX
80 Maiden Lane, Suite 704, New York, NY 10038

www.continuumbooks.com

All rights reserved. No part of this publication may be reproduced or transmitted in any form or by any means, electronic or mechanical, including photocopying, recording, or any information storage or retrieval system, without permission in writing from the publishers.

Copyright © Charlotte Methuen, 2008

Charlotte Methuen has asserted her right under the Copyright, Designs and Patents Act, 1988, to be identified as the Author of this work.

British Library Cataloguing-in-Publication Data
A catalogue record for this book is available from the British Library

Typeset by RefineCatch Limited, Bungay, Suffolk
Printed on acid-free paper in Great Britain by Biddles Ltd, King's Lynn, Norfolk

ISBN–13: HB: 978–0–567–03271–3
ISBN–10: HB: 0–567–03271–X

Contents

Acknowledgements — vii

1 Introduction — 1

Part I: Nature and Order in the Theology of the Reformers

2 Natural Order or Order of Nature? Natural and Moral Philosophy in the Thought of the Reformers — 7

3 *Lex naturae* and *ordo naturae* in the Thought of Philip Melanchthon — 19

Part II: Providence and the Interpretation of the Heavens

4 'This Comet or New Star': Theology and the Interpretation of the Nova of 1572 — 33

5 Special Providence and Sixteenth-Century Astronomical Observation — 48

6 Time Human or Time Divine: Theological Aspects in Opposing the Gregorian Calendar Reform — 61

Part III: Confession and Authority

7 From *sola scriptura* to *astronomia nova*: Authority, Accommodation and the Reform of Astronomy in the Work of Johannes Kepler — 77

8 On the Problem of Defining Lutheran Natural Philosophy — 94

Bibliography 112

Index 125

Acknowledgements

This book brings together a number of articles which further explore questions raised by my first book, *Kepler's Tübingen: Stimulus to a Theological Mathematics*. The debts incurred in the period of over ten years during which they were written are many and various. Particular thanks are due to Christoph Strohm, who encouraged me to deepen my thinking on Melanchthon's understanding of law; to Cees Leijenhorst and Christoph Lüthy, for their insights on natural philosophy; to Jürgen Hübner, for his willingness to discuss Kepler's theology; to John Brooke, for his invitation to participate in the European Science Foundation workshop programme on 'Religious Values and the Rise of Science in Europe', and for his continuing thoughtful advice; to Peter Harrison, for discussions about the emergence of the concept of laws of nature; to Judith Becker, for her willingness to read and comment on early drafts of much which appears here; to Thomas Kraft, who encouraged me to make this book a reality; to Timothy Bartel, for his intelligent and detailed copy-editing; and to Robert Franke, who has given so much support and encouragement along the way. A fellowship from the Alexander von Humboldt Foundation made possible the research which led to Chapters 4 and 5. Staff at the Staatsarchiv in Stuttgart, the Herzog August Bibliothek in Wolfenbüttel, the Historischer Lesesaal in Tübingen, the library of Christ Church College, Oxford, and the Duke Humphrey Library in Oxford's Bodleian have been unfailingly helpful in assisting me to track down obscure texts and references.

Thanks go also to the editors and publishers of the original versions of these articles for permission to reprint them in a revised form. Chapter 2 draws on 'Natural Order or Order of Nature? Natural and Moral Philosophy in the Thought of the Reformers', which appeared in *Studies in Science and Theology* 8 (2001/2), 57–75, reprinted by permission of Niels Henrik Gregersen and ESSSAT. Chapter 3, '*Lex naturae* and *ordo naturae* in the Thought of Philip Melanchthon', which appeared in *Reformation and Renaissance Review* 2.1 (June 2000), 110–25 (which was issue 3 of *RRR*), and Chapter 6, 'Time Human or Time Divine: Theological Aspects in

Opposing the Gregorian Calendar Reform', previously published in *RRR* 3.1–2 (December 2001), 36–50, are reproduced with permission of Ian Hazlett and the editorial board. Chapter 4, ' "This Comet or New Star": Theology and the Interpretation of the Nova of 1572', published in *Perspectives on Science* 5 (1997), 499–515, is reproduced with permission of Mordechai Feingold and the current publishers, MIT Press. Chapter 5, 'Special Providence and Sixteenth-Century Astronomical Observation, appeared in *Early Science and Medicine* 4 (1999), 99–113, and is reproduced by permission of the editors and Brill. Chapter 8, 'On the Problem of Defining Lutheran Natural Philosophy', which was first published in John Brooke and Ekmeleddin Ihsanoglu (eds), *Religious Values and the Rise of Science in Europe* (Istanbul: IRCICA, 2005), 63–80, is reprinted with permission of the editors. Chapter 7, 'From *sola scriptura* to *astronomia nova*: Authority, Accommodation and the Reform of Astronomy in the Work of Johannes Kepler', is published in English with the permission of Édouard Mehl, Alain Ségonds and Miguel Granada; it will appear in French in Miguel Granada and Édouard Mehl (eds), *'Nouveau Ciel, Nouvelle Terre': La révolution copernicienne dans l'Allemagne de la Réforme 1530–1630* (Paris: Éditions Les Belles Lettres, forthcoming 2008).

Whatever the deficits of this collection of essays – in some ways rather disparate – I hope that it will make a contribution to our deepening understanding of the complex relationship between theological impulses and observational science which helped to shape the beginnings of modern empirical science.

Charlotte Methuen
Oxford and Hanau
Advent 2007

1

Introduction

Modern discussions of the relationship between science and theology demonstrate increasing interest in investigating the ways in which the emerging natural sciences, rooted as they were in an intellectual culture which was dominated by theological concerns, were not only hindered but informed and stimulated by theology, and particularly by Protestant theology.[1] Recently, increasing attention has been paid to the ways in which humanist principles of Biblical interpretation, and the focus upon studying the original text, gave rise to an impulse to more precise observation of the natural world.[2] However, the impulse to study the natural world drew also upon the philosophical and theological assumptions about the structure of the natural world, and as Protestantism spread in the German-speaking lands of the sixteenth century these assumptions were naturally those which arose from the theology of the Reformers. For scholars shaken by the crisis of authority caused by the events of the Reformation, searching for a new route to certain knowledge, the doctrines of creation and of providence, with their fundamental assumption that God created and sustains the world for the good of its human inhabitants, could provide an impulse towards the

---

[1] The most recent discussions include: Funkenstein, *Theology and the Scientific Imagination*, Harrison, *The Bible, Protestantism, and the Rise of Natural Science*, Howell, *God's Two Books*, Lindberg and Numbers (eds), *God and Nature*, and Methuen, *Kepler's Tübingen*. For a discussion of the particular influence of Protestant theology on the rise of science, see, for instance, Harrison, *The Bible, Protestantism, and the Rise of Natural Science*, 5–8, and Charles Webster, 'Puritanism, Separatism, and Science'.

[2] See particularly Harrison, *The Bible, Protestantism, and the Rise of Natural Science*, Howell, *God's Two Books*, and Methuen, *Kepler's Tübingen*.

study of the natural world.³ The work of Philip Melanchthon, the 'Lutheran scholastic' who sought a new philosophical basis for understanding theology and in doing so laid strong emphasis on the study of the natural world, was particularly influential. Melanchthon did not always find himself in tune with the ideas of Martin Luther, and his approach was theoretical, in that he did not himself make observations of the natural world. Nonetheless, his approach to the natural world seems to have informed the ideas of those astronomers who argued that a strong conviction that God could be known and worshipped through the study of the natural world might support interpretations which contradicted the accepted Aristotelian or Ptolemaic worldview.⁴ Kusukawa has suggested that Melanchthon's approach amounted to a 'Lutheran natural philosophy',⁵ and although she has been careful to qualify this claim so as not to exclude a 'static' definition of Lutheranism,⁶ the question of confessional approaches to natural philosophy, and through it to empirical observation, remains open.

This book brings together a collection of articles previously published elsewhere which consider the question of the theological approach to and interpretation of the natural world. The first part begins with a brief examination of the place of the natural world, and of natural philosophy, in the theology of Martin Luther, Philip Melanchthon, Huldrych Zwingli and John Calvin, showing how different their approaches are, despite an underlying conviction that God has created the world to his own glory. Chapter 3 considers the relationship between natural law and the order of nature in the theology of Melanchthon, concluding that for Melanchthon the ability to understand number is a fundamental and innate principle which underlies the principle of order which is so central to his theology. Part II offers three studies of the interaction between theological argument and the interpretation of astronomical observations, with a particular focus on the arguments of a group of scholars centred on the South German duchy of Württemberg, and in particular its university, in Tübingen. Chapters 4 and 5 explore the role of the theology of providence in the interpretation of the nova of 1572, while Chapter 6 considers arguments against the proposed Gregorian reform of the calendar. The final part considers the implications of developments in astronomical observation for the theological foundation of Lutheranism. Chapter 7 explores Johannes Kepler's use of the doctrine of accommodation, its relationship to his understanding of

---

³ For a discussion of different theological formulations of the doctrine of providence and some of their implications for the study of the heavens, see Chapter 5.
⁴ As is the case, for instance, with Michael Maestlin. See Methuen, *Kepler's Tübingen*, 153–7, 171–7.
⁵ Kusukawa, *The Transformation of Natural Philosophy*.
⁶ Kusukawa, 'Lutheran uses of Aristotle', esp. 186–8.

hypothesis, and the way in which these shape his view of authority. For him, the heavens may almost be said to have an authority above that of Scripture. The concluding chapter discusses the extent to which it is possible to speak of a 'Lutheran' approach to natural philosophy or astronomical observation.

The nature of this work – a collection of essays – means that there is not a coherent argument being developed from chapter to chapter. Rather, this volume offers a series of snapshots and insights, which together nonetheless illuminate the interplay between theology and observation of the natural world in a Lutheran context during the later sixteenth century, and which emphasize both how difficult it is to generalize across Lutheranism and also how those who did wish to study the natural world could be served by the belief that the glory of God was revealed through what they saw.

Part I

Nature and Order in the Theology of the Reformers

2

Natural Order or Order of Nature? Natural and Moral Philosophy in the Thought of the Reformers

*I*n the late Middle Ages, but also in the sixteenth century, theology was the context in which philosophy took place. However, because of the structure of the curriculum, in which the study of the arts preceded the study of theology, philosophy was also the context of theology. Sixteenth-century debates about the nature and scope of philosophy shaped and were shaped by contemporary theological controversies and crises. During the sixteenth century, both theologians and philosophers redefined their understanding of human capabilities, whether through the acceptance of humanist views about how human beings could and had shaped their culture and their history,[1] or through the conviction of human sin and consequently human incapability. The interaction between these trends meant that sixteenth-century views of human nature and human potential to understand the world, particularly within Protestantism, were at once more optimistic and more pessimistic than those of their medieval predecessors. The attempts of Protestant theologians to reconcile these contradictory strands of thought had important consequences not only for their theology, but also for their understanding of natural philosophy and of ethics, and therefore for their understanding of the order of the world. This chapter considers the place of the natural world in the thought of Martin Luther, Philip Melanchthon,

---

[1] For a discussion, see Atkinson, *Inventing Inventors*, esp. 30–66.

Huldrych Zwingli, and (more briefly) John Calvin. It explores the ways in which a perceived natural order acted as a guiding principle in the understanding and portrayal of the natural world and society, focusing in particular on the question of what knowledge of God that order was thought to offer.

That nature and the universe were hierarchically ordered has been regarded as one of the fundamental principles of medieval and early modern theology and philosophy. Thus Allen maintains that

> the idea of a hierarchical order both in the heavens and on earth, in nature and in social relations and politics, was taken for granted well into the sixteenth century.[2]

In his biography of Calvin, William Bouwsma comments on the tension between the thought of Calvin as 'a philosopher, a rationalist and a schoolman in the high scholastic tradition', who 'craved desperately for intelligibility, order, certainty', and that of Calvin the 'rhetorician and humanist', who was rather 'inclined to celebrate the paradoxes and mystery at the heart of existence'.[3] The emphasis on order apparent in the former aspect of Calvin's thought is that highlighted by Christoph Strohm, who comments, in his study of the ethics of Lambert Danaeus:

> To the mental presuppositions which shaped the way in which people of the sixteenth century experienced their world, belongs the idea of an order which has hierarchical and complementary structure and which pervades the whole of creation. This God-given order defines that which is characteristic of the cosmos in contrast to unformed chaos and is the condition of all life. It guarantees the continuation and development of human life.[4]

The assumption that the world is divinely ordered according to physical and ethical laws pervades much of the philosophy and theology of the sixteenth century. However, as Schreiner has observed, that order is not invulnerable, and its vulnerability formed another important theme in sixteenth-century thought, interwoven with the first:

> The drastic effect of sin on the cosmos, the fragility of the universe, the need for a correct doctrine of providence, the remaining integrity of creation and human reason, and the revelation of God in the order and beauty of nature were themes expressed throughout the Reformation theology of the sixteenth century.[5]

---

[2] Allen, *Philosophy for Understanding Theology*, 161.
[3] Bouwsma, *John Calvin*, 230–1.
[4] Strohm, *Ethik im frühen Calvinismus*, 650 (translation mine).
[5] Schreiner, *The Theater of His Glory*, 119.

Sixteenth-century theologians sought to articulate the balance between the assumption of a fundamental order of the universe, including society, and the conviction that both sin and providence might intervene to destroy or disrupt it. Thus, whilst 'the revelation of God in the order and beauty of nature' is indeed a theme common to all the Reformers, they held a variety of views of the significance of the 'effect of sin on the cosmos', the 'fragility of the universe', and the post-lapsarian 'integrity of creation and human reason'.

The Reformers' attitudes towards the order of the world are most clearly revealed through the significance of moral and natural philosophy for theology, which generally reflects the weight given to natural theology and to natural law. Drawing on the creation narratives and wisdom Psalms of the Old Testament and from Paul's words in Romans 1.20,[6] all the Reformers regarded it as evident that God and God's providence were in some way and to some extent revealed in the natural world; however, they disagreed about what this natural theology could reveal about God. Similarly, human beings created *imago Dei* had an innate understanding of how a morally good life should be led – natural law – but this had been impaired by the Fall. The Reformers differed in their perception of what might be revealed through the natural world, and in the implications of natural law for moral life; they disagreed about the extent to which the perception of natural theology and of natural law had been distorted by sin and the Fall; and they had different understandings of how the natural world was to be defined.

Martin Luther

Luther taught that an awareness of the natural world leads to the praise of God.[7] Before the Fall, Adam and Eve, created *imago Dei*, lived in perfect knowledge of God, society and nature; Adam was a perfect philosopher, jurist and *medicus*. However, with the Fall this knowledge was lost.[8] Luther's theology must speak to a fallen world, which no longer has this original clarity of perception or perfection of knowledge, and therefore cannot rely on natural reason. For Luther, to place too great a dependence

---

[6] 'Ever since the creation of the world his eternal power and divine nature, invisible though they are, have been understood and seen through the things that he has made' (NRSV).

[7] For Luther's approach to nature, see Olsson, *Schöpfung, Vernunft und Gesetz in Luthers Theologie*, and Maaser, *Die schöpferische Kraft des Wortes*. For a brief summary, see Methuen, *Kepler's Tübingen*, 63–8.

[8] Luther, *In Genesin Enarrationum*, WA 42, 50, 90, 106, 153; LW 1, 66–7, 119–20, 141–2, 204, *Diui Pauli apostoli ad Romans epistola*, WA 56, 372.3; LW 25, 362–3; compare Olsson, *Schöpfung, Vernunft und Gesetz in Luthers Theologie*, 256–69; Schreiner, *The Theater of His Glory*, 115–16.

on human capabilities is to allow human reason to trespass onto the realms reserved for God. His main interest is so to delineate the boundaries of philosophical thought that the believer is aware of its limited application, and Luther is therefore less interested in exploring natural theology than in emphasizing that the study of natural or moral philosophy must not be allowed to seduce the student into believing that he has no need of God's salvation – that is, of the gospel.[9]

Luther was particularly (and vehemently) worried that the study of Aristotle's ethics had persuaded many people that they could themselves shape a moral life which would prepare them for salvation; moral philosophy must know its proper area of concern, which is civic, ethical life, and not questions of salvation. Nature demonstrates an order which makes it useful to human beings (especially in medicine and in astronomy, since the observation of the movements of the planets gives rise to knowledge about time, seasons, and so on), and Luther was aware of, and not uninterested in, this knowledge. However, his focus is precisely on the usefulness of nature; thus when discussing astronomy, he explicitly avoids any form of metaphysical explanation, either of the universe or of God.[10]

Luther does, however, believe that society should also demonstrate an order which he understands to be a natural hierarchy (e.g. the role of princes; or men, women and children in a descending hierarchy). This hierarchy and order is protected in part by (both political and natural) law and may be guided and guarded by ethical considerations. Moral philosophy can help in understanding this hierarchy, just as natural philosophy can help people to understand the order in nature. Implicit in Luther's thinking (albeit not explicitly stated) is the sense that natural philosophy and knowledge of the natural world is inferior to moral philosophy, which is in turn inferior to theology. But Luther is not really interested in establishing the relationship between different branches of philosophy; indeed he may even be said to have excised from his concept of natural law any understanding of a fundamental order of being,[11] effectively divorcing the moral order from the order of nature. Although Luther does believe there to be a natural order in the sense that both the natural world and society are ordered according to particular principles, he is not particularly interested in exploring how they might be linked through philosophical reasoning, in part because he does not understand them to be in any way informative about the salvific action of God, although he concludes that struggling to break

---

[9] See, for instance, Luther, *Galatervorlesung*, WA 40/1, 409–12; LW 26, 261–3.
[10] Maaser, 'Luther und die Naturwissenschaften', 25–41.
[11] This is the principal thesis of Maaser, *Die schöpferische Kraft des Wortes*; see esp. 237, 297–303.

out of one's place in the order of society is a sin,[12] as is a failure to take on the responsibilities of that place.[13] For Luther, the focus is upon the limit of what human reason can attain, a focus which tends to militate against his developing any suggestion that philosophical reasoning may be placed in an ascending hierarchy. Thus, he implies that the order of nature and the order of society are separate parts of the natural order; moral philosophy cannot be informed by natural philosophy, and natural philosophy in the sense of metaphysical speculation should not be undertaken at all.

When he drew upon philosophy at all, Luther inevitably (given the teaching of philosophy in the sixteenth century) appealed to Aristotelian philosophy, so that it is an Aristotelian order which permeates his understanding of how society is meant to be structured. In his focus on the restrictions of philosophy Luther stands in the late-medieval nominalist tradition. However, Luther is more radical in his conclusions: for instance, he was, and remained, adamant that philosophical explanations had no place in discussing theological mysteries such as the doctrine of the Eucharist.[14] Luther's strict separation between philosophy and theology meant that his focus was on the restrictions on human knowledge.

Philip Melanchthon

The situation is quite different in the case of Philip Melanchthon.[15] Melanchthon has a keen interest in the study of nature, and especially astronomy and medicine. The order which can be observed in the natural world both reveals the creative and providential aspects of God and offers a model for the order of society, and should serve to guide the observer to lead a life which is not only guided by the ethical principles but disciplined in prayer. For Melanchthon, the pre-Fall world was characterized by the perfect hierarchy of cause and effect, functioning by means of the stars and other heavenly bodies, which transmitted the will of God to the world, in particular through human beings. The Fall has caused disruption (and in

---

[12] *Admonition to Peace*, WA 18, 299–308; LW 46, 23–8.

[13] *Admonition to Peace*, WA 18, 293–9; LW 46, 19–23. Luther defines the responsibilities of princes in *Von weltlicher Uberkeytt*, WA 11, 271–8 (LW 45, 118–26).

[14] For Luther's attitude to philosophy, see Baur, 'Luther und die Philosophie', and compare Kopperi, 'Luthers Theologische Zielsetzung'. His reluctance to draw upon philosophical arguments in explaining theological concepts can be seen, for instance, in his rejection of the doctrine of transubstantiation in *De captivitate babylonica ecclesiae* (WA 6, 509–12; LW 36, 31–5).

[15] For these aspects of Melanchthon's thought, see Bellucci, *Science de la Nature et Réformation*; Frank, *Die theologische Philosophie Philipp Melanchthons*; Geyer, 'Welt und Mensch'; Kusukawa, *The Transformation of Natural Philosophy*. The following section is a summary of Methuen, 'The Role of the Heavens in the Thought of Philip Melanchthon'; see also *Kepler's Tübingen*, 68–105, and Chapter 3 below, where sources and further literature are given.

particular generation and corruption) in the sublunar world, but the heavens, composed of perfect quintessence, can still inform the observer of the will of God (as does, to an extent, the human body, since human beings have been created *imago Dei*). This is possible because even after the Fall the human mind possesses a 'natural light' given to it by God which enables it to perceive traces of God (*vestigia Dei*) in the observed order.[16] The soul remains in part *imago Dei*, and can be known better by applying the insights of medicine.[17]

The premises of both the order of nature (mathematical principles which allow the perception of order) and natural law (which Melanchthon identifies with the Decalogue and subsequently also with moral philosophy) are innate to human beings, set into their minds by the Creator as a reflection of the human status as the image of God. Indeed, these principles are related to each other:

> Just as in the theoretical sciences like mathematics there are certain general principles (*communia principia*) or κοινὰ ἐννόιαι ἢ προλήψεις, such as that the whole is greater than the parts, so too there are in the moral sciences also common principles and first proofs as rules of all human actions. . . . These can rightly be called natural laws.[18]

Just as the principles of mathematics or natural philosophy are related to the principles of moral philosophy or natural law, so too can God be known from the order which pervades not only the natural world but human society. In the final edition of his theological textbook, the *Loci communes*, and in the *Initia doctrinae physicae*, Melanchthon cites nine 'proofs' of the existence of God:

- from the order of nature, which could not have arisen by chance
- from the nature of the human mind, which could not have been created by a brute nature
- from the distinction between honesty and turpitude, and similar knowledge of nature, order or number
- from the truth of knowledge of nature
- from the human conscience
- from political society
- from the series of efficient causes, because it must have an end

---

[16] For the role of light in Melanchthon's thought, see Bellucci, *Science de la Nature et Réformation*, esp. 258–65 and 464–7; and compare also Schreiner, *The Theater of His Glory*, 119.

[17] As explored in Kusukawa, *The Transformation of Natural Philosophy*, 75–123.

[18] Melanchthon, *Loci communes 1521*, 100–3.

- from final causes
- from the interpretation of future events[19]

Here philosophical principles, the order of nature and the order of society occur side by side.

Although Melanchthon, like Luther, emphasizes that the knowledge attained through natural reason is not gospel and cannot lead to salvation, so that 'natural knowledge of God is knowledge of God's will towards sinners, not of his hidden nature',[20] Melanchthon nevertheless assigns a higher place to philosophy and connects philosophy and theology far more closely than does Luther.[21] For Melanchthon, the proper study of the natural world will lead to an understanding of moral philosophy, and to proper, pious behaviour, which can open the observer to God. He extends to other branches of learning, in particular natural philosophy and mathematics, the humanist idea that classical learning will lead to learned piety – *docta pietas*. Although Melanchthon recognizes that knowledge of God can only come from minds in which the seeds of this knowledge have been planted by God, his discussions of providence and of the natural order begin with knowledge of the world. In Melanchthon's thought, the order of nature must be taken to include the order of society and any philosophically defined order. Order is both a natural and a divine principle, which permeates the whole world in which we live, and knowledge of God, and of how God wishes society to be ordered, can be derived from the observation of this order.

Melanchthon's is an eclectic philosophy, based upon the premise that the human mind seeks – and moves towards – the truth. Drawing upon the theories, not only of Aristotle, but of the (natural) philosophers Plato, Cicero, Galen and Ptolemy in developing his theories,[22] he welds them together in a system based upon the principle of order. Melanchthon thus exemplifies the 'typical' emphasis on hierarchical order described by Allen, Bouwsma and Strohm. In his emphasis on order, his interest in natural law, and his concern to draw different philosophical systems into one, his thought bears a certain resemblance to that of the scholastics,[23] although

---

[19] Melanchthon, *Loci communes* (1543), CR 21, 641–3; and cf. Melanchthon, *Initia doctrinae physicae*, 200–2.

[20] Steinmetz, *Calvin in Context*, 27.

[21] This is less true of Melanchthon's earlier works, from 1520 until about 1535. From 1535 his theology, and particularly the structure of the *Loci communes*, originally based upon Paul's epistle to the Romans, becomes less biblical and draws increasingly upon themes from scholastic theology. Thus chapters on creation and providence appear first in the 1535 edition of the *Loci communes*.

[22] For Melanchthon's use of classical authors, see Bellucci, *Science de la Nature et Réformation*; Maurer, *Der junge Melanchthon*.

[23] This is true not only of Melanchthon's understanding of astronomy (see Maurer, *Der junge Melanchthon*, 1, 159–61) but of the presuppositions to Melanchthon's understanding of the law (see Chapter 3 below).

his belief that philosophy could help observers and scholars to understand God's created order and prepare them for an orderly life, his concern for education, and his encouragement of the teaching of natural and moral philosophy are clearly deeply influenced by humanist ideals.

Huldrych Zwingli

In his approach to discerning the structure of the world, Melanchthon begins (theoretically at least) with the observation of natural order, from which he believes the nature of God and God's intention for the world can be understood. In contrast, Zwingli's discussions of providence begin with the philosophical system by which he explains the world, and into which he incorporates God on all levels as an essential and active principle. Zwingli's presupposition is that the world cannot exist without the presence of God, so that 'if God were not (which is impossible), everything would cease to be'.[24] In his sermon on providence, Zwingli begins by arguing that God is the absolute good (*summum bonum*),[25] a principle which is arguably scriptural but which is also central to Aristotle's moral philosophy as presented in the *Nicomachean Ethics*.[26] Arguing from the identification of God with the absolute good, Zwingli asserts that 'providence must exist because the highest good necessarily cares for and orders all things'.[27] God is the essential, active principle of both the natural and moral worlds: 'what [Pliny] called nature, we call Deity';[28] that is, all things have their being in God. Indeed, 'God is for the universe what reason is to man'; all things are so done or disposed by the providence of God that nothing takes place without his will and command.[29] Zwingli understands God to be so immediate to creation that 'secondary causes are not properly called causes', but are rather to be understood as the result of God's direct intervention.[30]

Without faith, reason cannot bring about true knowledge of God, which is the enjoyment of God and the worship of God as the supreme

---

[24] Zwingli, *Sermonis de providentia Dei*, CR 93/3, 99 (*Schriften*, 4, 162).

[25] Zwingli, *Sermonis de providentia Dei*, CR 93/3, 70–8 (*Schriften*, 4, 141–6).

[26] Kraye, 'Moral Philosophy', 316. As she notes, during the Middle Ages and the Renaissance, 'Christians were in fundamental agreement that God was the source of man's ultimate happiness', regarded by Aristotle as the *summum bonum* (ibid., 316–17).

[27] Zwingli, *Sermonis de providentia Dei*, CR 93/3, 70 (*Schriften*, 4, 141), and compare Stephens, *The Theology of Huldrych Zwingli*, 93–4, and Schreiner, *The Theater of His Glory*, 117–18.

[28] Zwingli, *Sermonis de providentia Dei*, CR 93/3, 97–8 (*Schriften*, 4, 161).

[29] Zwingli, *Farrago annotationum in Genesim*, CR 100, 168, 179.

[30] Zwingli, *Sermonis de providentia Dei*, CR 93/3, 83, 111–13 (*Schriften*, 4, 151, 173–6). Presumably in criticism of the kind of Christian astrology taught by Melanchthon, Zwingli comments, 'we assign to the sun and the stars attributes which belong to God alone.... Their essence, power and activity is not their own but divine' (ibid., CR 93/3, 111–12; *Schriften*, 4, 174).

good.³¹ Indeed, true knowledge of God, including natural law (the worship of God and the love of neighbour), can be known only through the Spirit and not through reason.³² For, argues Zwingli,

> it is only the believer who sees God at work in all things in his power and providence. The unbeliever ascribes these things to man or nature, as if nature were something other than the power and providence of God.³³

Equally, without faith, the guiding principle may be perceived – as by the philosophers – but cannot be recognized as God.

Zwingli's understanding of God as a guiding principle of the natural world is complemented by his belief that although society's ills spring from the lack of true faith in God,³⁴ everything that takes place in society or in the course of human concerns is not a result of luck or coincidence but a direct result of God's providence.³⁵ There is little sense here of order as a guiding principle; indeed Zwingli explicitly warns against trusting any human sense of perceived order.³⁶ Indeed, the involvement of God in everything that happens in the world, and the arbitrariness of God's actions – for instance, his conviction that God intervenes in the ordered motions of the heavens in order to remind astronomers of their limitations³⁷ – indicate that Zwingli's understanding of providence focuses on special providence and ultimately on the unpredictability of the world.³⁸ Zwingli viewed natural and moral philosophy as servants of theology, but did not believe them to lead to reliable knowledge of God. Rather he emphasized that God can always intervene to set aside the order he has given to the world. Because his emphasis is on divine omnipotence rather than on divine order, Zwingli had little interest in establishing a hierarchy of disciplines such as that for which Melanchthon strove.

---

[31] Büsser, 'De Providentia', 15–16; cf. Stephens, *The Theology of Huldrych Zwingli*, 144.

[32] Stephens, *The Theology of Huldrych Zwingli*, 125–6 (esp. n. 70) and 140–1. The content of Zwingli's understanding of natural law (termed by Zwingli in German 'gsatz der natur' ('Gesetz der Natur'), which Stevens translates linguistically accurately, but somewhat confusingly, as 'law of nature') is discussed ibid., 131, 140–1, 300.

[33] Zwingli, *Farrago annotationum in Genesim*, CR 100, 10; cf. Stephens, *The Theology of Huldrych Zwingli*, 92.

[34] Stephens, *The Theology of Huldrych Zwingli*, 285.

[35] Zwingli, *Sermonis de providentia Dei*, CR 93/3, 202 (*Schriften*, 4, 251); cf. Büsser, 'De Providentia', 43.

[36] Zwingli, *Sermonis de providentia Dei*, CR 93/3, 194–5 (*Schriften*, 4, 246).

[37] Zwingli, *Sermonis de providentia Dei*, CR 93/3, 195–7 (*Schriften*, 4, 246–8), and cf. Büsser, 'De Providentia', 41–2.

[38] This is in stark contrast to Melanchthon, whose understanding of providence focuses on general providence. For a discussion of the application of providential theology to astronomical observation, see Chapters 4 and 5 below.

John Calvin

Calvin holds together aspects of Melanchthon's understanding of the order of nature with a strong sense of the unpredictability and unreliability of human knowledge. Like Melanchthon, he believes human beings to have been

> 'formed to be . . . spectator[s] of the created world' and endowed with eyes to see 'the world as a mirror or representation of invisible things' and so to be led through the contemplation of creatures to the praise of their divine author.[39]

However, this capacity has been damaged by the Fall, and is thus severely limited in its ability to perceive the divine through the natural world.[40] On the other hand, like Zwingli, Calvin's emphasis on God as almighty causes him 'to tie God's ever-present hand to each secondary cause and created movement' in the world of nature, and to emphasize 'the beautiful but precarious nature of creation'.[41] Thus while Calvin's arguments for the existence of providence are characterized by one concern – 'the attempt to find an indisputable foundation upon which to affirm that a reliable God controls a rational universe'[42] – and although God created a world characterized by order and harmony, order cannot be regarded as intrinsic to postlapsarian creation. Rather, the natural world has become altogether more dark and threatening: 'disorder [has] penetrated the physical elements . . . [and] the historical sphere is now also characterised by moral disorder'.[43] Despite the effects of the Fall, and 'in spite of natural and historical disasters',[44] providential order may nonetheless dimly be perceived through all the disorder. In human beings, created *imago Dei* with an ordered soul, the original order has also been defaced, and only a remnant remains. This remnant of conscience and natural reason is enough to contribute to the order and stability of the human realm and to allow society to continue a generally orderly course.[45]

---

[39] Steinmetz, *Calvin in Context*, 29, citing Calvin's commentary on Romans 1.18–21 (*Commentarius in Epistolam Pauli ad Romanos*, 28–31; *Commentary on Romans*, 29–32).

[40] Calvin, *Institutio christianae religionis*, I.5.

[41] Schreiner, *The Theater of His Glory*, 36, summarizing her arguments at 22–35. However, Calvin's attitude to secondary causes remains ambiguous (ibid., 30). Schreiner's study of the role of nature in Calvin's thought is the most detailed available, and this section is largely based upon it.

[42] Schreiner, *The Theater of His Glory*, 33.

[43] Schreiner, *The Theater of His Glory*, 33. Note that for Calvin, unlike Melanchthon, the Fall has consequences for the whole created universe, both heaven and earth.

[44] Schreiner, *The Theater of His Glory*, 33.

[45] Schreiner, *The Theater of His Glory*, 72, 94–5.

Calvin's approach to nature is in many ways characterized by the tension between his understanding of rational order and his perception of the effects of the Fall. This 'dynamic paradox', which, Bouwsma argues, permeates his thought,[46] manifests itself not only in Calvin's awareness, derived from the Stoics,[47] of the order which can be perceived in the divinely ruled world and of the relation between that order and the order of society, but also in his deeply pessimistic view of the effects of the Fall. The sustaining of the divine order on any level, let alone its perception, has to be understood as a victory of God over the forces of evil. Like Luther, Calvin is very conscious that he is living in – and writing for – a fallen world.

Conclusion

While the four Reformers considered in this chapter were agreed that the natural world proclaims the glory of God, the differences between them are marked. Luther's conviction of the priority of gospel over law renders him sceptical about what can be achieved in terms of knowledge of God through the pursuit of human arts, including natural philosophy. Melanchthon, on the other hand, although he is adamant that philosophy and human knowledge cannot bring about salvation, places a strong emphasis on order, such that he sees a close relationship between moral and natural philosophy, and regards the natural world, in particular the heavens, as a model for society (and indeed, for the Church). Melanchthon leaves his readers with a sense that natural order is related to divine order in such a way that God's providential actions can be revealed through the study of the order of the natural world and society. Zwingli's emphasis on God's ability and willingness to intervene in the running of the world reveals that his interest in providence focuses upon special providence. This restricts the value of natural philosophy, astronomy and other sciences, although Zwingli employs traditional natural and moral philosophical categories and explanations as a basis for his understanding of God's action. Calvin draws upon elements of all three, but his approach is characterized by a tension between divinely created order and the disastrous effects of the Fall. If the former is emphasized over the latter, the study of natural and moral philosophy becomes imperative and the two are likely to be closely related; if the latter is emphasized, the philosophical approach can be rejected as inapplicable to the fallen world.

---

[46] Bouwsma, *John Calvin*, esp. 230–4.
[47] See Strohm, *Ethik im frühen Calvinismus*, 128.

These differences are not confessional in any simple sense. The dispute between the adherents of Luther, who opposed the teaching of natural and moral philosophy in universities, and the followers of Melanchthon, who thought these subjects a necessary prerequisite to theology, dragged on at Lutheran universities until at least the end of the sixteenth century.[48] Similarly, the 'ordered' side of Calvin, which is close to Melanchthon's formulation, and the deep sense of order which they articulated here, certainly influenced Calvinist philosophy,[49] although Calvin's concern with the threatening, dangerous nature of the universe has also been important in the history of Calvinism. Zwingli's emphasis on God's ability to intervene in natural processes can be found in a somewhat modified form in the theology of some late-sixteenth-century Lutheran theologians, especially those opposed to the study of astronomy or of natural philosophy.[50]

The Reformers' approach to nature and society was shaped by their theological priorities and by their philosophical backgrounds and interests. The differences in their approach to the order of the natural world is a reminder of the complexity of the interactions between theology and philosophy in the formative years of Protestant theology.

---

[48] For the University of Tübingen, see Methuen, *Kepler's Tübingen*, 61–2. For Melanchthon's influence at other German universities, see Pozzo, 'Die Etablierung des naturwissenschaftlichen Unterrichts unter dem Einfluß Melanchthons'.

[49] Strohm argues, for instance, that the thought of Lambert Danaeus is shaped by the search for a theologically constitutive order: see Strohm, *Ethik im frühen Calvinismus*, esp. 594–608. Compare also the findings of Lohr, 'Latin Aristotelianism and the Seventeenth-Century Calvinist Theory of Scientific Method'.

[50] See Chapter 5 below.

3

Lex naturae *and* ordo naturae *in the Thought of Philip Melanchthon*

Of the Reformers discussed in the previous chapter, it is evident that Philip Melanchthon was the most interested in the study of nature; it is therefore unsurprising that he was the most influential in shaping that study. Melanchthon's particular focus on astronomy and natural philosophy or physics as a means for understanding and praising God, and his association with astronomers such as Casper Peucer (later his son-in-law), heightened his influence on those who observed the natural world — and has thus increased his significance in the eyes of historians of science.[1] His attempt to integrate theology and philosophy into a coherent system of thought is both a result of and a stimulus to his interest in the natural world as a source of divine knowledge. For Melanchthon, natural philosophy, moral philosophy and providential theology are closely related in a progression of sciences, and the ethical and providential explication of natural philosophy and astronomy acts as a linchpin holding his theological and philosophical systems together.[2] At the same time natural law, in the sense

---

[1] See, for instance, Westman, 'The Melanchthon Circle'.
[2] See particularly Kusukawa, *The Transformation of Natural Philosophy* and '*Aspectio divinorum operum*'. For Melanchthon's use of astronomy and mathematics, see also Methuen, 'Zur Bedeutung der Mathematik' and 'The Role of the Heavens'. Frank, *Die theologische Philosophie Philipp Melanchthons* offers a detailed consideration of Melanchthon's proofs for the existence of God and their foundation in cosmology and ethical systems. Geyer, 'Welt und Mensch' considers Melanchthon's understanding of the relationship between physics and ethics. Maurer, *Der junge Melanchthon* and Bellucci, *Science de la Nature et Réformation*, explore Melanchthon's philosophical background.

of innate ethical principles, also plays a central part in his ethical thought.[3] This chapter explores the relationship between these key areas of natural law and the order of nature in Melanchthon's thought. Since his discussions of natural law take place within the context of his discussions of law in general, the analysis which follows will also illuminate the role played by the order of nature in Melanchthon's understanding of the law. As a result of the connections Melanchthon himself makes between the theoretical principles of dialectics and the mathematical sciences and the order of nature, a further focus will be the clarification of the relationship between the *principia practica* and the *principia speculativa*.

The place of philosophy in Melanchthon's theology

Melanchthon's thought is generally recognized to demonstrate a shift of emphasis in the mid- to late 1520s, in part at least as a response to the disturbances in Wittenberg and other manifestations of spiritual power and political unrest. The guiding spirits of these movements understood their right to oppose the political hierarchy to be based upon an interpretation of Scripture which was not in line with Luther's own exegesis. Melanchthon's confrontation with them, coming as it did soon after his own conversion to evangelical ideas, led him to reconsider the relationship between faith and philosophy. From the late 1520s and the early 1530s, his writings demonstrate a renewed interest in the teaching of philosophy, together with a stress on the importance of education, and an emphasis on the centrality and necessity of ethics. Scholars have examined this focus on ethics from two different angles: in terms of the prominence of natural law as a central element in Melanchthon's theology,[4] and in relation to a growth in his interest in the teaching of natural philosophy, in particular anatomy,[5] and the mathematical sciences, especially astronomy.[6]

These aspects are not unrelated. Melanchthon's emphasis on teaching philosophy can be understood as forming the context for his response to the need for an ethical framework within which the gospel is to be preached and which offers an order essential to proper human life. Melanchthon was convinced that philosophy itself cannot lead an unbeliever to knowledge of God's salvific work – in this he is fundamentally committed to Luther's

---

[3] The role of natural law and its place in Melanchthon's thought is discussed by Frank, *Die theologische Philosophie Philipp Melanchthons*; see also Strohm, 'Zugänge zum Naturrecht bei Melanchthon'.
[4] See especially Strohm, 'Zugänge zum Naturrecht bei Melanchthon'.
[5] Kusukawa, '*Aspectio divinorum operum*'.
[6] See e.g. Methuen, 'The Role of the Heavens'.

conviction that this knowledge comes only through the gospel – but can only offer an understanding of the order which God wishes to be perceived in the world. As such, the study of philosophy belongs in Melanchthon's scheme of thought to the sphere of law, and not to gospel.[7] He nevertheless understands theology and philosophy to be intimately related: externally, the order of the natural world provides a route which may lead the observer to God,[8] whilst internally natural law is innate to human beings, set into their minds by the Creator as a reflection of the fact that human beings are created *imago Dei*.[9] These fundamental ethical principles are innate to the human mind, but so too is the understanding of number, which for Melanchthon underlies the perception of order. Indeed, for Melanchthon the understanding of number leads to the understanding of geometry, and thus to an appreciation of God's work (and even of God's mind) for, 'as Plato says, God always geometrizes'.[10] Like the human understanding of natural law, the human ability to appreciate mathematical, geometrical order makes it possible for human beings to recognize the order which God wants for the world, and to understand and judge their behaviour accordingly. It is therefore possible that although, as Strohm has argued, Melanchthon seeks to suppress the scholastic understanding that natural law is anchored in eternal law,[11] his appeal to innate mathematical premises in fact reveals a tacit belief that the basis of natural law is a form of 'eternal law' in the form of the created structure of the world. The challenge is to ascertain the way in which the premises of natural law and those of logic or mathematics fit together in Melanchthon's thought, and to investigate whether Melanchthon does in fact understand one of these sets of premises to be primary.

Melanchthon's definition of natural law

In his discussion of law in the first, 1521 version of the *Loci communes*, Melanchthon refers to three categories of the law: divine law (*lex divina*), natural law (*lex naturae*) and human law (*lex humana*). Divine law is made

---

[7] Melanchthon states this quite explicitly: see, for instance, *De discrimine evangelii et philosophiae*, CR 12, 690.
[8] This is particularly true of the heavens, but applies to the natural, sublunar world as well. See the discussions by Kusukawa, *The Transformation of Natural Philosophy*, esp. 100–7 and 160–73, Methuen, 'The Role of the Heavens', and Frank, *Die theologische Philosophie Philipp Melanchthons*, esp. 268–83.
[9] See particularly Strohm, 'Zugänge zum Naturrecht bei Melanchthon'.
[10] Melanchthon, *Praefatio in theoricae novae planetarum*, CR 2, 817. For the history of this phrase, said to have been written over the entrance to the Academy in Athens, see Ohly, 'Deus Geometra'.
[11] Strohm, 'Zugänge zum Naturrecht bei Melanchthon'.

up of those laws 'which God has placed in the canonical scriptures as a holy and unbreakable order', and has three parts: moral law (effectively identical with the Ten Commandments), judicial law and ceremonial law. Moral law is immutable, while judicial and ceremonial law are valid only in some ages and circumstances and have largely been superseded. Moral law, however, continues to apply.[12]

Melanchthon defines natural law to be 'common knowledge to which all people agree in the same way, to the extent that God implanted it into the heart of each of us with the intention of forming morals'.[13] Natural law, then, is the moral code acceptable to all human beings. Melanchthon argues that this is known to exist because of Paul's 'extraordinarily clever proof' at Romans 2.14–16: since the Gentiles have a conscience which either defends or condemns an action, there must be a law which is known also to them.[14] At this stage, Melanchthon defines three basic principles which make up natural law:

1. Honour God.
2. All people are born into society and should therefore not injure anyone.
3. Goods should be held in common.[15]

Melanchthon's argument for the existence of natural law is two-fold. Firstly, as has been seen, he argues from Romans 2.14–15. But Melanchthon also draws a parallel between natural law and the theoretical sciences:

> Just as in the theoretical sciences like mathematics there are certain common principles (*communia principia*) or κοινα ἐννόιαι ἤ προλήψεις, such as that the whole is greater than the parts, so too there are in the moral sciences also common principles and first proofs as rules of all human actions. . . . These can rightly be called natural laws.[16]

Melanchthon notes that Cicero, following Plato, understands the knowledge of natural law to be derived from human nature, a derivation which

---

[12] Melanchthon, *Loci communes 1521*, 110–11.
[13] Melanchthon, *Loci communes 1521*, 100–1.
[14] Melanchthon, *Loci communes 1521*, 100–1.
[15] Melanchthon, *Loci communes 1521*, 104–5. Melanchthon later lists four points, taking into account the effects of the Fall:

1. Honour God.
2. All people are born into society and should, therefore, not injure anyone.
3. As few injuries as possible should be allowed: people who breach the peace may be imprisoned.
4. Goods should be held in common. (ibid., 108–9)

[16] Melanchthon, *Loci communes 1521*, 100–3.

Melanchthon in 1521 finds unconvincing.[17] In the 1521 *Loci*, Melanchthon makes no explicit connections between moral philosophy and natural law. Nevertheless, even at this stage his conviction that natural law exists is founded upon the assumption that *principia practica* exist which are parallel to the *principia* which form the basis of theoretical sciences. But the parallel he draws does not place the *principia speculativa* within the realms of natural law, and he gives no indication of whence the theoretical principles might come.

Philosophy, law and mathematics

The basic structure of Melanchthon's definition of law remains the same throughout his career: the *lex Dei* is composed of three parts, the *lex divina* (of three types, of which only the *leges morales* are perpetually binding), the *lex naturae* and the *lex humana*. What changes, however, is Melanchthon's understanding of how natural law comes to be found in all human minds, and this is allied with an increasingly complex discussion of the relationship between natural law, the *leges morales* and the Decalogue on the one hand and moral philosophy on the other. Thus, in the 1535 edition of the *Loci*, Melanchthon discusses the relationship of philosophy and law. Under the heading 'lex divina' (somewhat confusing, since he refers here to the whole of what he generally calls *lex Dei*, rather than specifically to the *lex divina* as revealed by God in the scriptures), he explains the difference between human law, philosophy, and the law of God:

> Human laws merely require or prohibit external actions. Philosophy requires something besides, namely diligence not only in controlling external members but also in controlling the passions to a certain moderation and equality or agreeableness, and it is properly called ἠθικόν.... But the law of God does not only require or prohibit external actions, nor does it only require that diligence about which philosophers teach, namely the control of the passions, but it requires nature to consent to the will of God to the same measure as concupiscence is now repugnant to the law of God. And principally it requires firm knowledge of God, true fear of God, firm faith in God, perfect love.[18]

---

[17] Melanchthon, *Loci communes 1521*, 102–3. Melanchthon is probably referring here to Cicero, *De legibus*, 1, 515ff.; 6, 18ff. (ibid., 103 n. 264). For Melanchthon's use of Cicero in developing his theory of natural law, see Maurer, *Der junge Melanchthon*, 1, 207–9, and compare Wilhelm Dilthey, 'Das natürliche System der Geisteswissenschaften'.

[18] Melanchthon, *Loci communes* (1535), CR 21, 388–9.

Philosophy, in this case moral philosophy or ethics, requires the consent of the mind, while the law of God requires also the consent of the heart.

The law of God differs from the gospel in that, as Melanchthon argues in 1546,

> the law of God (*lex Dei*) is the teaching which tells us what kind of people we ought to be, and which works are laid down for God and for human beings, or is the teaching requiring perfect obedience to God; it is not the promise freely offering remission of sins, nor does it judge how we are to please God when we do not satisfy the law.[19]

Knowledge of the law allows the recognition of what perfect obedience to God might mean, although it can only be partially fulfilled because the Fall has spoiled the ability of the human heart or will to consent to the will of God. What ability human beings do have to assent to the will of God has been implanted in them by virtue of their status as creatures in the image of God.[20] Defining natural law in the 1535 *Loci*, Melanchthon notes:

> Natural law is the knowledge of divine law which has been set into human minds. Nor is anything in the whole nature of things better or more beautiful, nor is any vestige of God more present than the way in which God has impressed this his image and the likeness of his wisdom into human minds.[21]

In this discussion, Melanchthon assumes that the Decalogue is an expression of the fundamental principles of natural law.[22] As in the 1521 *Loci*, the Decalogue is identified with the *leges morales*, or unchanging part of the *lex divina*, which in turn is clearly restricted to the law as revealed in Scripture.[23]

The status of moral philosophy is rather less clear: despite the difference in the action of the will which Melanchthon identifies in the 1535 *Loci* as the essential distinction between law and (moral) philosophy, he is able to argue in 1546 that moral philosophy is part of divine law and the wisdom of God, precisely the same definition he applies to natural law.[24] As early as 1525 he defines natural law as 'opinions about morals which we hold from nature without learning',[25] and in 1529 he writes that 'ethical teaching is part of natural law'.[26] Although moral philosophy clearly requires learning,

---

[19] Melanchthon, *Philosophiae moralis epitomes*, CR 16, 22.
[20] Melanchthon, *Enarratio aliquot librorum Ethicorum Aristotelis*, CR 16, 279–80. Compare Strohm, 'Zugänge zum Naturrecht bei Melanchthon'.
[21] Melanchthon, *Loci communes* (1535), CR 21, 392.
[22] Melanchthon, *Loci communes* (1535), CR 21, 392.
[23] Melanchthon, *Loci communes* (1535), CR 21, 389.
[24] Melanchthon, *Philosophiae moralis epitomes*, CR 16, 23, and compare also 27.
[25] Melanchthon, *Argumentum et scholia in Officia Ciceronis*, CR 16, 636.
[26] Melanchthon, *Enarrationes aliquot librorum Ethicorum Aristotelis*, CR 16, 278.

and although technically it is 'that part of divine law which teaches about external actions',[27] the distinction between moral philosophy and natural law is increasingly being eroded. Because he sees the Decalogue as an expression of natural law, he can identify the first precept of natural law with the first tablet of the Decalogue – that God should be known and worshipped – while the second precept is summarized on the second tablet, in commandments governing human action and in particular civic life.[28] The first precept of natural law is thus identical with the primary concern of moral philosophy, namely the discovery of the *finis hominis*, which is 'to know God and to confess his glory', and to be obedient to God.[29] As Melanchthon asserts, 'nor is anything in the whole nature of things better or more beautiful, nor is any *vestigia Dei* more present than the way in which God has impressed this his image and likeness of his wisdom into human minds'.[30] Natural law is a consequence of the creation of human beings in God's image; although it is obscured by the consequences of the Fall the 'firm knowledge of the law of God, namely that God is the creator and governor of all creatures, that God is good and just, that God benefits the good and punishes the evil', together with the ability to differentiate between good and evil, remain the clearest vestige of the image of God.[31]

This clear association between moral philosophy and natural law appears to render natural philosophy and the other theoretical sciences irrelevant to the discussion. However, at around the same time that Melanchthon was arguing in the *Loci* for the privileged recognition of the *vestigiae Dei* in natural law, he was writing to Simon Grynaeus (in January 1535) arguing also that mathematics offers privileged access to the *vestigiae divinae*: the heavens offer 'vestiges of divinity' which can be unlocked through observations based upon geometrical and arithmetical techniques.[32] Implicit in this argument is (as has been noted above) his assumption that, 'as Plato says, God always geometrizes'. Because the human mind is made in the image of God, it also reflects aspects of the divine mind, particularly in its ability to recognize order and number.[33] Therefore, Melanchthon exhorts us to

> let superior minds, originating from the heavens, think about where they come from; from time to time let them study this theme and realize that this most beautiful spectacle of celestial bodies and movements

---

[27] Melanchthon, *Philosophiae moralis epitomes*, CR 16, 21. Note that the passions are here excluded.
[28] Melanchthon, *Loci communes* (1535), CR 21, prima tabula, 392; secunda tabula, 396.
[29] Melanchthon, *Philosophiae moralis epitomes*, CR 16, 28.
[30] Melanchthon, *Loci communes* (1535), CR 21, 398–9.
[31] Melanchthon, *Loci communes* (1535), CR 21, 398–9, and compare also 367–70.
[32] Melanchthon, *Praefatio in theoricae novae planetarum*, CR 2, 814–21.
[33] Melanchthon, *Praefatio in arithmeticen*, CR 11, 290–1.

has not been set forth to humankind in vain, and let them enquire into the order of these most admirable things, because it is most appropriate to human nature and because it carries great usefulness for life.... For as in all things it is best to be ruled by God, so in this consideration of studies, as we might call it whenever we contemplate the sky itself, let us be reminded of the Architect. Let us not think that he instituted this amazing order and transmitted the understanding of these movements to the human race for nothing.[34]

The *vestigiae divinae* are thus not only a part of natural law, but are also to be found in the natural (or, in this case, the celestial) world; the human ability to recognize them is implanted by God. The mathematical sciences, in particular astronomy, and natural philosophy, in particular the knowledge of plants and of anatomy, are important ways into this knowledge of God, and the recognition of God's attributes which results from them gives rise to the same kind of knowledge which is to be expected from the proper understanding of natural law:

There was an important reason why God gave these testimonies about himself. For, since we have learnt from them that God is the ruler of all, we understand that we must obey him and we acknowledge that from him comes order both in our own minds and in wider society, and that there are penalties consequent on upsetting this order.... In short, this order of heavenly laws tells us much about God and about how to behave, testifying that all these things have been founded for the sake of the human race.[35]

From the order of nature, as from natural law, then, the observer is meant to learn that God exists and that society should be ordered according to certain principles of justice and injustice. Indeed, the point of studying natural philosophy, Melanchthon explains in the first, 1549 edition of the *Initia doctrinae physicae*, is that it leads the observer to know God.[36] Melanchthon distinguishes natural philosophy from moral philosophy in that the former seeks out the *causae hominis* while the latter deals with the *finis hominis*.[37] Ultimately, both lead to praise of the Creator, for God is both the cause and the end of humankind.

Melanchthon roots his understanding of the human capability to recognize the *vestigiae divinae* in the status of human beings as created in the image of God. However, in his discussions of natural philosophy and

---

[34] Melanchthon, *De astronomia et geographia*, CR 11, 294.
[35] Melanchthon, *De astronomia et geographia*, CR 11, 297.
[36] Melanchthon, *Initia doctrinae physicae*, CR 13, 181.
[37] Melanchthon, *Philosophiae moralis epitomes*, CR 16, 28.

astronomy, Melanchthon seems to accept that the heavens will in fact yield a more precise knowledge of God as creator and sustainer of the world than will human society or observations of the sublunar world, since these latter have been affected by the Fall while the heavens have not.[38] Indeed he seems to suggest that it is the human ability to recognize the *principia practica*, and, more to the point, to have the will to obey them, which has been most severely impaired by the Fall; the ability to recognize the *principia speculativa* and to reason logically is to a large extent preserved.[39] The fallen state of the human mind and soul to some extent impairs their ability to understand the natural world; for instance, as he argues in 1545, the fallen human intellect is unable to interpret directly the motions and light of the stars.[40] Nevertheless, the impression remains that Melanchthon believes the *principia speculativa*, that is, the principles of mathematics and of dialectics, to be less susceptible to the results of the Fall than are the *principia practica*. This impression is strengthened by his conviction that the *principia practica* must be understood in parallel to the *principia speculativa*, and by his use of arguments from mathematics and natural philosophy to justify this. Thus, in the 1535 *Loci*, he argues:

> Just as light is sent excellently into the eyes, so certain *notitiae* are set into the human mind, or a certain light by which they know and judge things. The philosophers call this light the knowledge of principles; they call it κοινάς ἐννοίας and προλήψεις. And just as they teach about the *principia speculativa* that people know their nature and understand [them to be] certain, so it is also to be known about the *principia practica*, that is, about natural law.[41]

Similarly, in the 1559 *Loci* he compares the eternal nature of natural law to eternal mathematical truths such as that twice four is eight.[42] Nonetheless, the status of the *principia speculativa* and of the knowledge acquired by means of them remains unclear, for his comparison of them to the *principia practica* places them outside the system of natural law, and his definitions of divine law and human law also seem to exclude them.

Dilthey has suggested that, for Melanchthon, physics is the point at which it is possible to cross from the axioms of the theoretical sciences to the truths of life, and that his consciousness of God is rooted in both the ordered universe and in the human moral sense.[43] Given Melanchthon's

---

[38] As argued in Methuen, 'The Role of the Heavens'.
[39] Melanchthon, *Loci communes* (1535), CR 21, 399–401.
[40] Melanchthon, *Praefatio in libros de iudiciis navitatum*, CR 5, 822–3.
[41] Melanchthon, *Loci communes* (1535), CR 21, 392.
[42] Melanchthon, *Loci communes* (1535), CR 21, 711–12.
[43] Dilthey, 'Das natürliche System der Geisteswissenschaften', 177–9.

teleological interpretation of natural philosophy and astronomy, the obvious context for a consideration of the knowledge of God which comes through observation of the natural (or celestial) world is in the discussion of the first precept of natural law, namely that God should be known and worshipped, and in Melanchthon's later works he does indeed begin to move towards an argument that the order of the natural world must be seen as part of the first natural law. In 1550, he explains that

> Just as this most beautiful order – the positions of the bodies, of the heavens, of the air, of the earth surrounded by the oceans – and like the order of the motions of the heavens – which brings about the changes of time, days and nights, summer and winter – has been shaped and is preserved by God, so too the whole of this political order – the bond of marriage, empires, the distinction of states, contracts, judgements, punishments, indeed all most true statutes – originate from God.[44]

It is this human, rather than the natural, order which witnesses to the Decalogue and the correct human rule which is a reflection of the presence of natural law in human hearts.[45] In the *Loci* of 1559, however, Melanchthon notes that the first of the natural laws is that (among other things) 'the human mind should know that God is the eternal mind, the conditioner of good things, and that God is to be obeyed. . . . There are many demonstrations of these laws', including the '*pulcherrimus ordo naturae*' demonstrated by the principles of physics.[46] The proofs of the existence of God given in the 1559 *Loci* include both the order of nature and the order of society.[47] Through the order of nature the primary task of natural law can thus be fulfilled: recognition of God's creation and providential action.

At the same time, however, the parallels that Melanchthon draws between the principles of the theoretical sciences and the principles of natural law place the former outside the realm of natural law. Melanchthon seems simply to assume that theoretical principles exist and are self-evident. Indeed, knowledge of them offers evidence that innate principles do exist, and is thus evidence for the possibility of natural law. Indeed, as has been noted above, Melanchthon argues (following Plato) that mathematical systems are imprinted on the human mind because the mind is number, and that this recognition of order is an attribute which human beings receive from God, since they are created such that their human minds reflect God's

---

[44] Melanchthon, *De legibus*, CR 11, 912.
[45] Melanchthon, *De legibus*, CR 11, 912.
[46] Melanchthon, *Enarratio aliquot librorum Ethicorum Aristotelis*, CR 16, 386.
[47] Melanchthon, *Loci communes* (1559), CR 21, 641–3, see pp. 12–13 above.

mind. Melanchthon's constant reiteration that order is paramount seems to depend on a fundamental conviction that numerical order is both fundamental to the structure of the world and more apparent than ethical, social order. Thus in 1550, Melanchthon explains that

> there are many testimonies of God impressed into this whole heavenly system (*machina*) and all the bodies of the world which testify that this world was not created by chance but was shaped and is preserved with wonderful skill by an architectonic mind. But the clearest testimony is the way in which the eternal mind transmitted into human beings its light and its wisdom, which teach us that God is a comprehending mind, ordering numbers and the grade of things, free, distinguishing actions, discerning good and evil, good, equitable, true, guarding order, horribly opposed to all confusion and disorder.[48]

Order is primary, even though the concept of order is not mentioned in Melanchthon's definition of law.

There are certain parallels between Melanchthon's discussion of the relationship between moral and natural philosophy in the context of law and his use of ethical examples in his discussions of dialectics and syllogistic logic.[49] Here his central concern seems to be for 'certainty in instruction and for the method by which such certainty can be achieved',[50] and in the *Erotemata dialectices* Melanchthon argues that the (innate) principles, syllogistic reasoning and experience are the criteria for certain knowledge.[51] The appeal to innate principles in this context, taken together with Melanchthon's interest in relating moral and natural philosophy on the basis of tacit assumptions which seem to reflect the scholastic understanding of eternal law, suggests Melanchthon's intention or need to anchor his intellectual system in an order which is more than human. Thus, despite that fact that Melanchthon's definition of natural law may be argued to be subjective and anthropomorphic,[52] he does in fact seek to root his definition in a system which transcends the limits of human knowledge. For Melanchthon, God is fundamentally orderly; the divine is that which brings order into the world.

---

[48] Melanchthon, *De legibus*, CR 11, 908.
[49] As discussed by Kusukawa, '*Vinculum concordiae*', esp. 342–7. For similar themes in Melanchthon's approach to rhetoric, see Maurer, *Der junge Melanchthon*, 1, 171–214.
[50] Kusukawa, '*Vinculum concordiae*', 346.
[51] Melanchthon, *Erotemata dialectices*, CR 13, 647.
[52] Strohm, 'Zugänge zum Naturrecht bei Melanchthon', 341–7 (Strohm entitles this section of his article 'Subjektivisierung und Anthropozentrisierung').

God, law and the order of the world

Melanchthon's portrayal of divine order doubtless draws heavily upon the philosophy of Plato and of the Stoics.[53] But at the same time, it is very close to the Augustinian and particularly the Thomist understanding of *lex aeterna* as a divine, ordering principle which permeates the world and its doings.[54] Augustine's doctrine of *lex aeterna* 'seeks to describe the unity behind the multiplicity of things; it depends upon the order and peace of creation; it defines creation's beauty, its *pulchrum esse*'.[55] Similarly, Thomas Aquinas constructs a hierarchy in which eternal law offers the framework within which other forms of law have their place. Clearly included in Aquinas' concept of *lex aeterna* are the order and measure by which God has shaped the world,[56] and the implication is that knowledge of the world, in the form of knowledge of order and shape, is not only available to human beings but will lead them to a better understanding of God. Although Melanchthon does not draw explicitly upon the concept of *lex aeterna* – indeed, his definition of (divine) law restricts law to a biblical expression – the scholastic *lex aeterna* nevertheless seems to permeate his understanding of the theological implications of knowledge of the natural world.

This tacit assumption of the existence of the *lex aeterna* permeates Melanchthon's understanding of the way that God has shaped the world and uses it to instruct humankind. Created in the image and likeness of God, the human mind, even in its fallen state, reflects the mind of God in its understanding of mathematical order and its grasp of the principles of natural law. For Melanchthon, God stands for order – indeed, in a fundamental sense, God *is* the ordering principle in the world – and the capability to perceive and grasp this order is one of the greatest gifts which God has given to human beings.[57]

---

[53] See particularly Maurer, *Der junge Melanchthon*, 1, 129–70, and Bellucci, *Science de la Nature et Réformation*, 129–217.

[54] For the scholastic understanding of natural law and its relation to *lex aeterna*, see Fellermeier, 'Das Naturrecht in der Scholastik', and compare also Wolf, 'Zur Frage des Naturrechts'.

[55] Wolf, 'Zur Frage des Naturrechts', 183.

[56] Fellermeier, 'Das Naturrecht in der Scholastik', 340–2.

[57] While he would defend God's right to overturn the normal running of the world, Melanchthon has no patience with the idea that the primacy of the *potentia absoluta* could be used to render futile the study of the order of the world. Not all sixteenth-century theologians associated order with the divine in this way. As discussed in Chapter 2, others, including Luther, pointed to the unpredictable nature of God's interventions in the world and were less inclined to see order in the world as a signpost to the creator God. For Luther's approach to the natural world and its place in his theology, see Olsson, *Schöpfung, Vernunft und Gesetz in Luthers Theologie*, and compare Maaser, *Die schöpferische Kraft des Wortes*.

Part II

Providence and the Interpretation of the Heavens

4

'This Comet or New Star': Theology and the Interpretation of the Nova of 1572

In November 1572, a bright light, which is now known to have been a nova, appeared in the night sky. It was visible throughout northern Europe, a cause of fear to many who saw it and a cause of attention and concern to astronomers and astrologers alike. The Danish astronomer Tycho Brahe studied the star, puzzled over its status and provenance, and began to gather the opinions of other observers.[1] The English astronomer John Dee turned his attention to the phenomenon. Throughout Germany, dukes and princes summoned their astronomers and astrologers and demanded an explanation of the appearance. Astronomers responded with reports and treatises. In Tübingen, the seat of the university of the southern German duchy of Württemberg, the phenomenon was observed by the professor of mathematics, Philip Apian, and by his student Michael Maestlin, who was later to become the teacher of Johannes Kepler. Caspar Peucer, mathematician at the University of Wittenberg, made a series of detailed observations available to the Elector of Saxony. Cyprian Leovitus, astronomer to the Palatine, wrote a brief treatise. Wilhelm IV, Landgrave of Hesse, a noted stellar observer, had initially been too busy with state affairs to notice the new appearance in the heavens, but once his attention had been drawn to it, he too began to compile observations.[2] These and other reports passed from one court to the next, and from one astronomer to another. A

---

[1] He collected these opinions in his *Astronomiae Instauratae Progymnasmatum*, finally published in 1602 (Brahe, *Opera Omnia*, vol. 3); see Hellman, *The Comet of 1577*, 111–17.
[2] Moran, 'Wilhelm IV of Hesse-Kassel', 85–6.

number eventually reached Tycho Brahe, who incorporated them into his study of the new star. The concern of princes and other state leaders to understand the meaning of the star – interpreted both astronomically and astrologically – promoted the exchange of information and furthered the advance of astronomical understanding.[3]

This chapter will consider one small exchange in the broader flow of information: the letters and observational reports which passed between Wilhelm IV, Landgrave of Hesse, and Ludwig, Duke of Württemberg, in the winter of 1572/3. This correspondence offers a good example of the way in which aristocratic interest (not to mention princely competition) could foster discussion of astronomical theories, but it also provides a fine cross-section of the range of conclusions which were drawn from these observations and of the arguments which led to these conclusions. The primary aim of this chapter is to assess the way in which astronomers appealed to different disciplines in interpreting the observations they made in the winter of 1572/3, drawing upon their understanding not only of mathematics and astronomy but also of natural philosophy and especially of theology. The exchange of views between Hesse and Württemberg, and the observations and interpretations which were cited in the course of this exchange, use arguments drawn from a range of different disciplines and arrive at a range of different conclusions. In doing so, they offer both a corrective to the too easy assumption that 'better' observations or instrumentation automatically lead to 'better' results or to 'better' interpretations of those observations,[4] and a reminder that, as Peter Barker has pointed out, it is unwise 'to treat science as an activity in which most of the time things go right'.[5] Even given the undoubted limitations on the freedom of interpretation and of expression possible in such a correspondence between astronomers and their patrons, this exchange of observations and interpretations offers a useful insight into the processes by which sixteenth-century observers assessed new phenomena. Moreover, it demonstrates the care needed to understand the terminology by which they sought to describe their conclusions.

The correspondence on the 'new star'

The correspondence between Ludwig of Württemberg and Wilhelm IV of Hesse concerning the new star opened at the end of December 1572, when

---

[3] As discussed by Moran, 'Wilhelm IV of Hesse-Kassel'; idem, *Patronage and Institutions*; Westman, 'The Astronomer's Role'.

[4] The terms 'better' and 'correct' in quotation marks will refer to those sixteenth-century observations and conclusions which concur with those of twentieth-century astronomers.

[5] Barker, 'Copernicus, the Orbs, and the Equant', 322.

Philip Apian, Professor of Astronomy and Mathematics at the University of Tübingen, sent a report of his conclusions about 'the new star or comet' from Tübingen to Ludwig in Stuttgart. Apian's report seems to have been written in response to a (lost) request by Ludwig for information. Ludwig forwarded a copy of Apian's report to his cousin Wilhelm in Kassel.[6] Wilhelm responded initially with a letter dated 14 January 1573, enclosing a brief summary of his own observations together with a manuscript copy of a treatise sent to him by Cyprian Leovitus.[7] Two days later Wilhelm wrote again to Ludwig, sending him copies of Caspar Peucer's observations and conclusions, and discussing Apian's report in greater detail.[8] In February, Ludwig wrote at least twice to Wilhelm, although only somewhat illegible drafts of these letters survive in Stuttgart. Finally, on 26 February, Apian sent to Ludwig his own response to Wilhelm's opinions, also enclosing a *Iudicium* on the new star by 'an author who prefers to remain unnamed'. At some stage, probably in late January, Ludwig had received a further report on the new star by the astronomer Samuel Eisenmenger.[9] The correspondence also includes a further anonymous mathematical exposition of the '*nova stella*', apparently an early version of Michael Maestlin's *Demonstratio astronomica loci stellae novae* (1573).[10] Of these eight correspondents, only seven actually contributed opinions to the debate. Although Ludwig seems to have acted as the catalyst for the discussion, his letters express only his wish to know what is being observed, his interest in promoting the debate, and his desire to reach a decision about how the observations should be interpreted. Ludwig was not a 'prince practitioner' in the mould of Wilhelm of Hesse, but even under the Regency of his mother, the young Duke took a keen interest in the university of Tübingen, its professors and

---

[6] The correspondence is held in the Hauptstaatsarchiv, Stuttgart (hereafter HStAS), A274 Bü 21. The individual letters and reports are unnumbered, and will be referred to by author (and addressee, in the case of letters) and date. A shorter version of Apian's letter to Ludwig, in Latin and addressed to Wilhelm IV of Hessen can be found in Brahe, *Astronomiae Instauratae Progymnasmatum, Opera Omnia*, 3, 158–61.

[7] Leovitus himself later wrote to Ludwig, enclosing printed but unbound copies of his treatise on the new star in both Latin and German.

[8] The correspondence with Peucer includes copies of two letters from Peucer, the first dated 9 December 1572 and the second January 1573. The latter can be found in Brahe, *Astronomiae Instauratae Progymnasmatum, Opera Omnia*, 3, 120–3.

[9] The correspondence has been kept roughly in reverse chronological order, with some letters in two copies, not necessarily kept together. Thus Eisenmenger's report, dated 7 January 1572, is to be found between two copies of the report by Cyprian Leovitus, and forwarded to Ludwig by Wilhelm with a letter dated 14 January 1572.

[10] This exposition is neither dated nor attributed; however, the text and diagrams coincide exactly with Maestlin's text. I have dated the manuscript to 1572, since it opens with a reference to 'anno domini 1572' rather than to 'anno superiori 1572' as in the published version (Maestlin, *Demonstratio astronomica*, 27); however, a date of early 1573 might also be possible. My attribution and dating have been confirmed by Granada, 'Michael Maestlin and the New Star of 1572', who offers a careful comparison of the texts, discussing the development in Maestlin's ideas.

its teaching.[11] Ludwig's ultimate responsibility for the spiritual and economic well-being of his people brought with it the task of overseeing the censorship of books and opinions. Throughout his reign, Ludwig was to call for *Iudicia* to be written on new astronomical works as well as astronomical phenomena,[12] but although he encouraged discussion and facilitated the exchange of information, Ludwig was not himself an observer.

Describing the 'new star': the terminology and its meaning

Observers of the 1572 nova had a variety of opinions about what they could be observing. Aristotelian natural philosophy taught that there could be no substantial changes in the heavens, and for observers who adhered to this teaching, any new appearance in the skies generally had to be viewed as a comet, meteor or similar phenomenon occurring in the sublunar sphere. The majority of astronomers, astrologers and mathematicians who observed the new star probably took this view.[13] Increasingly, however, as observational methods improved, astronomers were coming to realize that, Aristotelian philosophy notwithstanding, comets and other phenomena could and did appear above the moon.[14] All the reports which are collected in the Hesse–Württemberg correspondence, except that of Cyprian Leovitus, which does not discuss the question, agree that their observations place the new celestial object[15] further from the earth than the moon. All

---

[11] Ludwig was here following in the footsteps both of his father, Christoph, and also of his grandfather, Ulrich, who had taken measures to reform the University of Tübingen and shaped it to promote his theological and political aims (discussed in Methuen, 'Securing the Reformation through Education'). Ludwig was closely involved in the appointment of the university's professors, overturning the university's own decision on more than one occasion. The copious correspondence concerning Nicodemus Frischlin, a controversial poetics professor at the university, demonstrates the way in which Ludwig could be drawn into university disputes (HStAS, A274 Bü 45, 46, 47).

[12] See, for instance, Maestlin's *Iudicium* (1586) on Frischlin's work (discussed in Methuen, *Kepler's Tübingen*, 115–17), and the associated correspondence (HStAS, A274 Bü 46, # #19, 26, 28). Ludwig also requested Maestlin to provide an opinion on the 'heavenly appearances' observed in Prague in 1596, and another on Johannes Kepler's work, *De natura et significationibus cometarum* (correspondence in HStAS, A202, Bü 2551). Westman, 'The Astronomer's Role', has shown that this advisory role was common for sixteenth-century university as well as court astronomers, and Schöner has investigated the role of professors of mathematics at the University of Ingolstadt, showing that they were responsible to the university, the Duke and the Arts Faculty (*Mathematik und Astronomie*, 285–313).

[13] See Hellman, *The Comet of 1577*, 115–16; eadem, 'A Poem on the Occasion of the Nova of 1572', 306–9; also Busch, *Von dem Cometen*; Graminaeus, *Erklerung oder Auslegung eines Cometen*.

[14] For a discussion of cometary theory prior to 1572, see Hellman, *The Comet of 1577*, 13–117, and compare Barker, 'The Optical Theory of Comets'.

[15] I use this imprecise term deliberately to reflect the uncertainty of the contemporary interpreters about what it was they were observing.

agree that it is very bright, brighter than any other star or even than the planets, and that, like a star, it twinkles. All observe that over the course of several weeks, both the colour and the size of the object change noticeably. But they are not agreed about precisely what it is that they are observing.

In retrospect, from the perspective of modern knowledge of the nova, it appears that the central factor in assessing the conclusions of sixteenth-century observers must be their understanding that what they were seeing was a new star. For observers in the sixteenth century, however, or at least for those who contributed to the correspondence discussed here, the question tended to be phrased somewhat differently. For Ludwig, for instance, the burning issue which gave rise to the correspondence was to ascertain whether or not the object was a comet,[16] although in the later letters he consistently refers to it as a star.[17] However, not all observers asked the question 'What is this object?', and even when they did, they did not invariably phrase the question as 'Is this a comet or a (new) star?' Although this formulation of the question was indeed central for some of the observers who contributed to the Hesse–Württemberg correspondence, for others it evidently appeared not only relatively unimportant but almost tautological. The terms 'comet' and 'star', clearly distinct in modern terminology, could in the sixteenth century refer to the same celestial object. Thus, Cyprian Leovitus refers in the sub-title of his printed report to the 'comet or new star', but in the body of his treatise sometimes to 'the star' and sometimes to 'the comet'.[18] Similarly, in his first letter, Wilhelm states explicitly that 'we do not find this star . . . to be a meteor or a comet', but goes on to refer to it initially as a comet, concluding finally that it is in fact a '*stella portentosa*', and referring thereafter to it as a star.[19] This terminological ambiguity means that even when an observer concludes that the 'comet or new star' is either a comet (in the case of Philip Apian) or a star (the conclusion of, for instance, Wilhelm of Hesse and Caspar Peucer) this classification is less clear than the term used might make it seem. A star may be any celestial object above the moon, including the planets, while a comet may or may not be supra-lunar, depending on the author's position with regard to Aristotelian cometary theory. The terms used, and thus the conclusions

---

[16] Ludwig to Wilhelm, 29 December 1572, fo. 1ʳ: 'Und nachdem wir nit wissen mögen ob es ein Comet oder anders [illegible word: possibly "meteor" but certainly not "stern"] auch deswegen und etlichen uf mancherlej weiss dans geredt und gehalten worden. So haben wir darüber bej dem astronomico unser hohen Shul zu Tübingen bericht einnemen lassen.'

[17] Ludwig to Wilhelm, draft letter 5 February 1573; also Ludwig to Wilhelm, 14 February 1573.

[18] Leovitus, *Von dem neuen Stern* and idem, *De nova stella*, and see esp. the marginal notes. Moreover, in one manuscript copy of his report, the title refers only to 'the comet' while the report refers only to 'the star', but this seems to be a copyist's error (Cyprianus Leovitus, 7 January 1573, comparison of the two copies in the Stuttgart archives).

[19] Wilhelm to Ludwig, 14 January 1573; cf. Wilhelm to Ludwig, 16 January 1573.

drawn by these observers, can only be understood by considering the meaning attributed by each observer to the terms he uses, and this in turn can only be discerned from a detailed examination of his arguments.

Interpreting the 'new star'

Before returning to the question of the meaning of the terminology, it is, however, important to note the relative unimportance of the question 'What are we observing?' for three of the seven commentators involved in this correspondence. The reports of Cyprian Leovitus and Samuel Eisenmenger, and the brief anonymous comment forwarded to Ludwig by Philip Apian, contain little or no observational evidence and virtually no discussion of what the object might be. As has already been noted, Cyprian Leovitus refers to the object interchangeably as a star and a comet. He explains its existence by asserting that it has 'been set alight by Jupiter and Mars'[20], thus following one of Aristotle's explanations for the appearance of (sub-lunar) comets.[21] Although Leovitus does not discuss its distance from the earth, he notes that the object does not move against the sphere of fixed stars, and that it behaves like a star, circling the pole star like other stars, but, unlike them, is bright enough to be visible even in daylight.[22] However, his belief that it has been caused by Jupiter and Mars appears to place it closer to the earth than the fixed stars. Leovitus makes no attempt to resolve this apparent ambiguity, and perhaps did not recognize it as such: apparently he understands the term 'star' to refer to any celestial (supra-lunar) object, and does not insist that a comet must be sublunar. The principal concern of his treatise is not observation or explanation of observation, but interpretation, in the sense that he seeks to understand the meaning of the celestial object for the world. Arguing from the appearance of similar stars 'in the same place in the sky' in the years 945 and 1245, and drawing on the long-established association between such stars and their 'strange and ominous effects', Leovitus warns of the terrible results which this star will have for the world, including war, confusion in civil and religious affairs and a long, hard winter.[23]

---

[20] Leovitus, *Von dem neuen Stern*, fo. Aii^r.

[21] Aristotle taught that comets may be formed in two ways: either when the emissions of the earth condense below the lunar sphere and ignite, or when the influence of a celestial body causes a comet to generate in the fiery sphere (Aristotle, *Meteorologia*, 344a5–345a10; compare Barker and Goldstein, 'The Role of Comets', 304–5, although their reference in n. 11 seems to be incorrect).

[22] Leovitus, *Von dem neuen Stern*, fo. Aii^{r-v}.

[23] Leovitus, *Von dem neuen Stern*, fos Aii^v–Aiii^r. The association of such concerns with comets was very common: for similar reactions to the comet of 1531, see Kusukawa, *The Transformation of Natural Philosophy*, 125–6. Heerbrand predicted that the comet of 1577/8 heralded war, famine and inflation (Heerbrand, *Predig von dem erschrockenlichen Wunderzeichen am Himel*, 4).

The report of Samuel Eisenmenger and the brief anonymous note differ from that of Leovitus in that both accept that the object is a new star and do not use the term 'comet' to describe it. Their central concerns are, however, very similar. Eisenmenger comments that although a new star has never been seen from the beginning of the world, this object would appear to be one, since it is above Jupiter and Venus and brighter than anything else in the sky. Although Eisenmenger refers to the object consistently as a star, he offers no further discussion of his use of the term; neither does he discuss in detail whether the object is a star or something else. He does, however, suggest that the object is 'of the same nature as Jupiter and Venus', implying that he understands it not to belong to the sphere of the fixed stars, but to be one of the wandering stars, or planets.[24] Like Leovitus, Eisenmenger's principal concern is to assess its meaning, 'for according to the learned opinions of astrologers and theologians such stars are portents of the future'. Here he turns to an assessment of the biblical evidence that such a star will precede the end of the world, considering the possibility that this is a diabolical portent, but concluding that God has sent the star to bring people to faith in Christ.[25]

The third of these reports, that by an anonymous author, comments that 'some say it is a comet but others disagree', but 'all are agreed that it is far above the moon and because of this do not doubt that it is in the eighth sphere among the fixed stars'.[26] It has no motion except the motion of the fixed stars, and its colour has changed from the colour of Venus to the colour of Mars. However, for the anonymous author, as for Leovitus and Eisenmenger, the question of its distance from the earth is peripheral: more important are the interpretation of the star and the question whether it heralds the end of the world. This author believes that it does, regarding the star as presaging the kind of apocalyptic battles described in Ezekiel (chapters 38 and 39) and Revelation (chapter 20), and warning against the papal forces and the incursions of the Turks.[27]

The 'new star' as a message from God

For all these authors, the question of the correct identification of the object is secondary to that of its interpretation as a portent of the future and to an understanding of it as a message from God. This wider, interpretative

---

[24] Eisenmenger, 'De Stella Nova', fo. 1ᵛ.
[25] Eisenmenger, 'De Stella Nova', fo. 2ʳ.
[26] Anonymous, 'Iudicium', fo. 1ʳ.
[27] Anonymous, 'Iudicium', fos 1ᵛ–2ʳ.

concern is typical for its time;[28] it is not, however, confined to the three reports which lack observational detail. It would be incorrect to conclude that more exact observers were not concerned with such matters. With the single exception of Michael Maestlin, all the authors whose opinions are gathered in this correspondence address this question, and all come to what is essentially the same conclusion: the object, whether it is a new star or a comet, whether natural or supernatural, has been sent by God as a message and warning for humankind. Such a concern is entirely congruent with the Christian astrology taught by Philip Melanchthon and other theologians; not only did this form the framework within which many observational astronomers conducted their work, but it could also act as an incentive to more precise observation, on the grounds that these would yield a better understanding of God.[29] However, as will be seen from what follows, the relationship between an observer's theological interests and the conclusions he draws from his observations is not always simply that of an impulse to exact observation.

In contrast to the first three reports, Philip Apian's is very concerned with the making of accurate observations which will enable the distance of the object from the earth to be established. Nevertheless, the arguments which he puts forward in the discussion of whether or not this object is a comet include other, non-observational factors. Evidence for its being a comet, in Apian's opinion, are the conditions under which it has appeared: its appearance in the heavens was preceded by a major conjunction of the planets Jupiter and Saturn, a celestial event which, in Apian's opinion, is often associated with the subsequent appearance of a comet. Moreover, this winter is very cold, and, like Leovitus, Apian believes that such weather often results from the appearance of a comet. On the other hand, observations show that this object behaves in many ways like a star: it twinkles, a sure indicator that it is very distant; it has no observable motion relative to the fixed stars; and its parallax places it far beyond the moon, probably in the sphere of the fixed stars.[30] Apian does not think, however, that this

---

[28] For an instance of such concern in Zurich, see Hellman, who suggests that this 'unusual celestial phenomenon was useful for the Protestant reformers' ('A Poem on the Occasion of the Nova of 1572', 308). All the correspondents discussed here were not only Protestant but Lutheran. However, the star excited interest among Catholics too, and at least one Catholic interpreter understood the 'comet' as an omen for the downfall of Luther's followers (this is argued, for instance, by Graminaeus, *Erklerung oder Auslegung eines Cometen*).

[29] For a discussion of this motivation in the work of Copernicus, see Wrightsman, 'The Legitimation of Scientific Belief', esp. 55–62. For theology as a motivation for astronomy in the thought of Philip Melanchthon, see Kusukawa, *The Transformation of Natural Philosophy*, 124–44, and Methuen, 'The Role of the Heavens'. For the theological impulse to study of the natural world in the work of Kepler and of his teachers, see Hübner, *Die Theologie Johannes Keplers*, and Methuen, *Kepler's Tübingen*, esp. 107–58.

[30] Apian to Ludwig, 26 December 1572, fos 1ᵛ–2ᵛ; cf. Brahe, *Astronomiae Instauratae Progymnasmatum, Opera Omnia*, 3, 158–9.

evidence is sufficient to allow the conclusion that this object is a star. He disagrees with the *physici* and other defenders of Aristotle's theory of comets, and does not accept that comets are flaming coagulations of sublunar exhalations; nor does he believe that comets ever occur below the moon. But he does believe that there are different types of comets. Others have argued that this object cannot be a comet because it has no tail, citing the opinion and observations of Apian's father, Peter Apian, that a comet always has a tail which always points away from the sun.[31] Philip Apian, however, is of the opinion that comets do not necessarily have tails, and he cites observations by Pliny and Ptolemy to support this view. His own conclusion is that what he is observing is a tail-less comet, a '*stella secunda*' or 'secondary star' which has been caused by a conjunction between Jupiter and Saturn. Although it occurs in the sphere of the fixed stars, he concludes that the object is neither a true fixed star nor a planet as, he writes, 'I believe no experienced [astronomer] will dispute.'[32] Apian concludes that the object is not a true star, although he does not know how to define it and in fact continues to refer to it as 'the comet or new star'.

Naming or defining this object is, however, less important to Apian than settling the question whether this is a natural, as opposed to a supernatural or miraculous, phenomenon, and it is in this context that he turns to a discussion of the object's significance. Again, he declines to give a final opinion on this matter, explaining that only when the object has disappeared from the sky will it be possible to place it in the context of eclipses and conjunctions, which is necessary if it is to be interpreted. However, he is prepared to assert that

> the comet or star has been created as a warning by Almighty God, because before he sends sickness and punishment, the merciful Father usually sends a good preacher to the pious and to the Godless, which is intended to convince and comfort the pious in the fear of God and remind the wicked and the Godless that they should live a contrite life if they wish to avoid punishment.[33]

As such, it must be seen as a sign from God, and since it is known that 'whenever comets have appeared, strange and sad events follow, as can be seen from the reports and observations of many historians and astronomers', such events can be expected in the future.[34] However, this still does not

---

[31] For a discussion of the development of this argument during the sixteenth century, see Barker, 'The Optical Theory of Comets', 6–10.

[32] Apian to Ludwig, 26 December 1572, fos 2ᵛ–3ʳ; cf. Brahe, *Astronomiae Instauratae Progymnasmatum, Opera Omnia*, 3, 159–60.

[33] Apian to Ludwig, 26 December 1572, fo. 5ʳ⁻ᵛ.

[34] Apian to Ludwig, 26 December 1572, fo. 4ʳ.

preclude the possibility that the 'comet or new star' is natural. Since Apian believes that all comets appear above the moon, he has no difficulty in concluding that this message from God has come about through the normal courses of the natural world.

Despite Apian's conviction that no astronomer could disagree with him, both his observations and his conclusions were rejected by a number of other experienced astronomers. Wilhelm of Hesse agreed neither with Apian's calculations of the distance of the object from the earth, nor that the object they were observing was a comet, nor that it was a natural phenomenon. According to Wilhelm's observations and his measurements of parallax (measurements and observations made, as he notes with pride, with instruments far superior to those available to Apian), together with his subsequent calculations, the object is indeed above the moon, but it is not in the sphere of the fixed stars. Instead, Wilhelm's observations place it 'in the upper sphere of Venus', which suggests that it has something in common with the planets.[35] Wilhelm criticizes Apian's suggestion that the object has been caused by a planetary conjunction as an overly astrological and misguided attempt to portray the appearance of this object as part of the natural course of the universe. For himself, Wilhelm is convinced that 'this star is something supernatural, since from certain and careful observations we could not find that it is a meteor or a comet, nor that it was constituted in the elementary region'.[36]

In a second letter, Wilhelm turns to Apian's report, criticizing his conclusions, and particularly the suggestion that the object observed is a comet. Although Wilhelm does not agree with Peter Apian's finding that a comet's tail always points away from the sun (since, he claims, the tail of the comet of 1555 pointed in the other direction), he is persuaded that a comet should have a tail. Therefore, since this object has no tail, it cannot

---

[35] Wilhelm to Ludwig, 14 January 1573, fo. 2^{r-v}.

[36] Wilhelm to Ludwig, 14 January 1573, fos 1^r–2^v: 'Und sovil das iudicium betrifft, seint wir mit euch aus angetzognen vrsachen vnd motiuen durchaus einig, vnd habens alzeitt darfur gehalten, wie wir Ihnen am erstenn gesehenn, das dieser Stern aliquid supernaturale sei, Dan niemall wir aus gewissen vnd vleissigen observationibus, et considerationibus nicht befinden konnen, das er ein meteoron oder Cometa sei, noch in elementum regione constituirt.' In Wilhelm's opinion, the differences between his conclusions and those of Philip Apian are to be found in the fact that Apian was lacking in instruments – probably, he writes scathingly to Ludwig, because the University of Tübingen 'is too poor to buy good mathematical instruments' (Wilhelm to Ludwig, 14 January 1573, fo. 1^v). In a later letter, Philip Apian responds scathingly to this suggestion: although he had enjoyed reading Wilhelm's report and observations, Apian could only repeat that his own observations were correct, and he protested that there was no lack of instruments in Tübingen (Apian to Ludwig, 26 February 1573). Given the investment of time, money and expertise which Wilhelm put into developing and improving the accuracy of his instrumentation (as described by Moran, 'Princes, Machines and the Valuation of Precision') it is doubtless true that his instruments were superior to those available to Apian at Tübingen. Nevertheless, in this case it was Apian's observations which were the more accurate.

be a comet.[37] Moreover, Wilhelm inclines to the belief that change can only take place naturally in the sublunar sphere. The only possible explanation of this new phenomenon is, therefore, that it is a miracle, and a sign put in place by God. Since this means that the appearance of this object is a supernatural event, it then becomes quite possible for Wilhelm to understand this object as a 'star of portent', albeit a 'star' which is placed not with the fixed stars but within the sphere of Venus.

Theology and the interpretation of the 'new star'

Although Wilhelm's measurements of the distance of the nova from the earth were 'incorrect', his perception that it is above the moon is 'correct'. This, taken together with his 'correct' conviction that the object cannot be a comet and with his belief that only God can introduce non-cometary change into the heavens, leads him to the 'correct' conclusion that he is observing a new star. However, his definition of the term 'star' allows him to place the object in the sphere of Venus, associating it with the wandering, rather than with the fixed, stars. This conclusion is only possible because of Wilhelm's theological conviction that although God normally allows the world and the heavens to take their course according to the laws of nature (part of general providence), God may also choose to intervene in this course and override those laws (special providence). In this case, Wilhelm understands God's intervention to be of such moment that he writes 'to our beloved cousin and relative the Kurfürst that we believe this to be a sign that we should conduct ourselves as if in the Last Days'.[38] As a miracle, the new star must be of particular import. However, in this case Wilhelm's theological understanding is not simply an aid to understanding the star's meaning; his identification of it as a star is informed by theological as well as astronomical and mathematical insights, and this combination of factors precludes any possibility that the so-called star might be understood to be a fixed star.

A similar process of argument can be observed in the reports of

[37] Wilhelm to Ludwig, 16 January 1573, fo. 1ᵛ: 'Was aber des Appiani Judicium betrifft, das ist mehr Astrologicum et Phijsicum, quam Astronomicum, das er aber bricht [?] das die Cometen alzeitt Ihre Comas ex obiectu Solis sollen extendiren, wie sein vatter solchs zu demonstriren unnderstanden, das mag wol zu Zeitten geschehen, es geschieht aber nicht allzeitt, wie solchs der Comet so Anno [15]55 erscheinen, der seijne Comas gerade im Contrarium extendirt, klar aussgewiesen. Darumb halten wirs nochmals beij unns mahr pro portento, dann pro Meteoro oder naturali apparentia.'

[38] Wilhelm to Ludwig, 14 January 1573, fo. 1ʳ: 'Darumb vnd dieweil er in aetherea regione Comparairt, da dan die phijsici keiner generation odder corruption stadt geben, so ist er bei uns dero wunderbarlicher, haben auch darumb sobalt wir Inen gesehen, vnd eher wir euer Judicium bekommen, vnsern freundtlichen lieben Vetter vnd Schwagern, den Khurfursten geschrieben, das wir inen fur der Zeichen, eins so vorm Jungsten tage hergesehen, solten halten.'

Caspar Peucer, for whom the identification of the celestial object also poses a major problem. Like Wilhelm, Peucer believes that past observations of comets demonstrate that this object cannot be a comet since it has no tail. Moreover, Peucer states explicitly that the principles of physics show that there can be no changes in the supra-lunar region, that is, outside the elementary sphere. For such a change to be natural, therefore, it must be sublunar. However, Peucer's observations of the object's distance from the earth place it outside the sublunar sphere, and, indeed, much further from the earth than the sphere of Venus. For Peucer, as for Wilhelm, the only possible explanation is that God has chosen to establish a new star in the heavens, an event which must be counted among the miracles.[39] Like Wilhelm, Peucer's theological conviction that God can transcend the laws of nature allows him to identify the appearance for what it is – a new star – and thereby to override Aristotelian restrictions on the possibility of change above the moon. Unlike Wilhelm, however, Peucer places the new star above all the planets and notes both that it twinkles and that it moves like the other fixed stars. Peucer's understanding of the new star thus places it in the sphere of the fixed stars and attributes to it the characteristics of other fixed stars, despite its changing colour and size. Although his understanding of Aristotelian natural philosophy is contradicted by his observations, his theological understanding of the possibilities of God's special providence allows him to reconcile the two and to arrive at a 'correct' interpretation of what he has observed.

Maestlin's observations lead him to conclude that the new star is placed in the 'eighth sphere or firmament of fixed stars'.[40] His argument differs, however, from that of Apian or that of Wilhelm and Peucer. Like Apian, Maestlin notes that the question whether the observed object is a comet or a star has not been settled. However, in contrast to his teacher, Maestlin tends towards the conclusion that it is a star. Since, like his teacher, he accepts that comets do not exist below the moon, so that the supra-lunar sphere admits change naturally, he has no need for recourse to divine intervention to explain this appearance. However, unlike Apian, Maestlin seems to accept that comets must have tails. This tail-less object cannot, therefore, be a comet and Maestlin seems happy to conclude that it is a fixed star.[41]

---

[39] Peucer to the Elector of Saxony, 9 December 1572 (copy), fo. 3^r: 'Quare interea agnoscimus et fateamur, singulare Dei opus et inter miracula referendum esse, quae praeter et extra nature ordinam citra opem secundarem causarum Deus excitat, et proponi diuinitus statuamus, ut uelut πρόδρομος preco ante filium Dei, fatales et ultimas mutationes generi portando exuscitet et reuocet homines ad poenitantium.' Peucer's letter here contradicts Hellman's assertion that he believed the star to be below the moon (Hellman, *The Comet of 1577*, 115).

[40] Maestlin, 'Noua stella', fo. 1^r; cf. Maestlin, *Demonstratio astronomica*, 28.

[41] Maestlin, 'Noua stella', fo. 1^r; cf. Maestlin, *Demonstratio astronomica*, 28.

Of the discussions of the 1572 nova which make up the Hesse–Württemberg correspondence, only Maestlin's is exclusively mathematical.[42] Peucer's reports include extremely detailed tables of observations, and Wilhelm and Apian's reports show that they were also seeking to make accurate observations. Ironically, of the three of them it was Wilhelm, with his superior instrumentation, who in fact produced the least 'correct' interpretation of the nova. Nevertheless, all of these observers sought to make use of the best possible observations in their attempt to understand what they were observing. However, of the seven observers who contributed to the correspondence, only Maestlin and Peucer concluded that the celestial object was a star in the sphere of fixed stars. A superficial reading of the conclusions of Eisenmenger, Wilhelm of Hesse and the anonymous observer, all of whom use the term 'star', would suggest that they shared the 'correct' interpretation of Maestlin and Peucer, but in fact they all placed the 'star' lower than the fixed stars. Apian, on the other hand, placed the object 'correctly' in the sphere of the fixed stars, but his flexible use of the term 'comet', combined with his conviction that comets were supra-lunar, led him to classify it as a comet, or '*stella secunda*': his terminology tends to disguise the fact that his observations sited the object closer to the 'correct' placing of the star than did those who called it a star but placed it in the lower spheres. Such ambiguities of astronomical terminology in the definition of stars, comets and planets continued well into the seventeenth century; indeed, Ariew has argued that discussions about comets and old and new stars grew more rather than less ambiguous, and distinctions between them were increasingly difficult.[43]

Of equal interest here is the role played by theology in the interpretative process. As has been noted above, theology has relatively often been identified as a motivation for making better observations, and Maestlin was later to use a biblical justification for the exact study of the heavens to justify his critique of the Aristotelian theory of comets.[44] However, although such attempts to make exact observations sprang from this motivation, theology played a different role in the arguments of Peucer and of Wilhelm. Here, theology was an intrinsic part of the argument which enabled these convinced Aristotelian observers to acknowledge and explain what they had observed. The conviction that God not only ruled the world but intervened

---

[42] The longer, published version includes a discussion of other factors, including theological and astrological considerations, but these do not appear in this early version. For further differences, see Granada, 'Michael Maestlin and the New Star of 1572'.

[43] Ariew, 'Theory of Comets', esp. 363–7.

[44] Maestlin, *Observatio et demonstratio cometae*, and idem, *Consideratio et observatio cometae*; for a discussion of the theological arguments used here, see Methuen, *Kepler's Tübingen*, 152–8, 173–82.

in its running allowed these astronomers to accept observations which were theoretically impossible according to natural philosophy as being the results of supernatural, divine events. Without their confidence in God's ability to create new stars at will, Peucer and Wilhelm could not have argued for the setting aside of Aristotle's theory of the immutability of the heavens. Of course, this theology could also have a stifling effect on astronomy and natural philosophy, and was used by some theologians and philosophers to illustrate the futility of any systematic study of nature.[45] In this case, however, it was the possibility of such a radical intervention by God which enabled Peucer and Wilhelm to accept and interpret their observations. Their conclusion was that the new star they had observed must have been supernatural. Theology served as an authority which allowed accepted principles of natural philosophy to be discarded without any loss of faith in the unity of the world. Although the clear distinction made by Wilhelm and Peucer removed the new star from the realm of the natural, their approach nevertheless assumed that their study of the heavens would reveal the will of God. This was the shared assumption of all natural philosophers until Newton and beyond,[46] and it provided the motivation for all the observers in the Hesse–Württemberg correspondence, whether they saw the new star as a natural or a supernatural phenomenon. With the exception of Maestlin, all these observers interpreted the object not just as a warning, but explicitly as a warning sent by God (and indeed Maestlin also subscribed to this view in the published version of his work). It was precisely because astronomy was done in order to know God that theological insights about the ways in which God worked in the world could and did inform observers' interpretations of their observations.

For those involved in the Hesse–Württemberg correspondence, the general consensus was to be that the object which had appeared in the sky was indeed some kind of new star. Even in Tübingen, Apian's personal opinion that it was a comet or '*stella secunda*' notwithstanding, this definition passed into accepted parlance; Nicodemus Frischlin published a work on the appearance and meaning of the 'new star',[47] and later in the century preachers were to refer to the 'new star' of 1572 as a messenger from God.[48] As the Hesse–Württemberg correspondence demonstrates, the use of this

---

[45] See, for instance, Andreae, *Christliche, notwendige vnd ernstliche Erinnerung*; for discussion see Methuen, *Kepler's Tübingen*, 125–8, and compare also Chapter 5 below.

[46] See Cunningham, 'How the *Principia* Got its Name', and compare Harrison, 'Newtonian Science, Miracles, and the Laws of Nature'.

[47] Frischlin, *Consideratio nouae stellae*.

[48] See for instance Jakob Heerbrand's sermon 'on the terrible miraculous sign in the heavens, the new comet . . .' (*Ein Predig von dem erschrockenlichen Wunderzeichen am Himel*).

terminology does not necessarily imply that observers saw this 'star' as a star in the modern sense. But whatever they did see, it was not only their astronomical observations, but also their theological convictions that allowed the observers of the nova to describe what they saw.

5

Special Providence and Sixteenth-Century Astronomical Observation

*I*t has become apparent that many sixteenth-century observers undertook the study of the natural world to the glory of God, believing that they would gain a better understanding of God through the knowledge gained from their endeavours. However, the theological basis for such an interpretation of the observation of the natural world could take various doctrinal forms, which were not mutually exclusive, and all of which were rooted in the belief that God created the world. For example, it could express itself in terms of the doctrine of providence, or in the wisdom tradition of the Old Testament, or in the understanding that the natural world may be understood as the *liber naturae*, and thus as a source of divine revelation in some way parallel to the Bible.

Each of these positions could lead to a theological justification of observation which was stated in terms of the glorification of God, and thus the fact that observers were drawing on different theological traditions may seem to be of little importance. But in fact, the differences between theological traditions have implications for their treatment of nature. While both the theological metaphor of the *liber naturae* and (perhaps to a lesser extent) the Old Testament wisdom theology focus upon the order which is to be found in the created world, the doctrine of providence, with its emphasis on God's continuing care for the world, not only understands God to have established and to sustain the order of the normal course of the world (general providence), but also allows the possibility of God's intervention in the normal course of the world (special providence). Providence is thus concerned not only with the order which may be perceived in nature, but with the ways in which that order may be changed by God. This chapter explores the role of the doctrine of providence in the responses to the 1572 nova considered in Chapter 4. It will first investigate whether there is a difference in emphasis in the doctrine of providence adopted by those theologians who are favourable to the study of the natural world compared to the doctrine of those who are not, and second, it will consider the approach of observational astronomers to phenomena they regarded as events of special providence.

These enquiries offer a historical investigation of Whitehead's assertion that 'faith in the possibility of science, generated antecedently to the development of modern scientific theory, is an unconscious derivative from medieval theology'.[1] Even into the seventeenth century, the work of most natural philosophers seems to be predicated on the assumption that an order might be imposed upon nature, and many historians have assumed in their turn that the concept of an ordered world was derived from, and continuous with, medieval understandings of order in the world.[2] However, John Brooke has argued that this assumption of continuity is one of several rather general theses which have plagued the portrayal of the relationship between science and religion, and that it must be treated with caution. It may therefore be useful to investigate the extent to which theologians, in discussing aspects of the doctrines of creation and providence which might be derived from observation of the natural world, did in fact assume that creation was ordered, and to what extent God was believed to be able to intervene in that order.

The distinction between the normal, ordered running of the world and the possibility of divine intervention in that order was retained by natural philosophers and scientists well into the seventeenth century.[3] However, between 1500 and 1700 a range of phenomena which had previously been regarded as results of God's intervention in the world came to be understood as part of the normal order of the world.[4] It will be argued in this chapter that such a shift involved a change in the understanding of the order of the world, and consequently a reconsideration of the boundary between general and special providence. The doctrine of providence is thus central in demonstrating some of the criteria involved in making this shift.

---

[1] Whitehead, *Science and the Modern World*, 19 (as quoted in Kuhn, *The Copernican Revolution*, 122). Compare also Wallace, 'Translator's Preface', in Blumenberg, *The Genesis of the Copernican World*, p. xi.

[2] Brooke, *Science and Religion*, 4.

[3] For a discussion of this distinction and its implications in the work of seventeenth-century scientists, see Brooke, *Science and Religion*, 117–51, and compare Brooke, 'Science and Theology in the Enlightenment'.

[4] Comets would be one example.

The doctrine of providence and the order of creation

The question of God's intervention in the normal order of providence emerged in the Middle Ages, probably in the wake of the reception of Aristotelian philosophy by Thomas Aquinas.[5] Aquinas believed that, while the structure of the universe has a basic order, God may choose to intervene and to change the established order or the normal running of the universe, because he could have chosen to establish another order. God may also choose to act through secondary causes,[6] and may thus break into the normal chain of causality. However, Aquinas also thought that such divine interventions are rather to be disregarded, because the basic order of the universe is more important.[7]

Aquinas here indicates a distinction which in sixteenth-century Lutheran theology was given a terminological basis: the difference between *providentia generalis* and *providentia specialis*.[8] The distinction is itself clearly older than the sixteenth century, but it was in that century, and particularly in the thought of the forerunners of Lutheran orthodoxy, that it gained importance. Whilst it arose in the context of discussions of the relationship between the *potentia Dei absoluta* and the *potentia Dei ordinata*, as is clear from Aquinas's treatment of it, the distinction between general providence and special providence is different from that between the *potentia Dei absoluta* and the *potentia Dei ordinata*. The latter distinction is concerned with the arbitrary nature of the order of this universe, with the sort of order God might have chosen to create had he created not this world with its order but some other in its stead, and with the status of the order in such potential universes. However, the terms *potentia Dei absoluta* and *potentia Dei ordinata* do not describe anything so simple as the world which God chose to create as opposed to that which God might have chosen to create, and although later medieval thinkers laid more emphasis on the order of the world as it was created, 'even those thinkers who stressed the perfection and order of our world believed that the choice of God to actualize *this* order was unaccountable and arbitrary'.[9] Discussions of the

---

[5] Bernhardt, *Was heißt 'Handeln Gottes'?*, and compare also Cunningham and French, *Before Science*, esp. 185–97.

[6] Aquinas affirms that God gives things 'natures and powers with which to act': *Summa contra Gentiles*, 3, 71. See also Cunningham and French, *Before Science*, 191–2.

[7] Aquinas, *Summa Theologica*, 1, 103 and 1, 105, 6 (ST 14, 2–33 and 78–83).

[8] I have not found these terms used earlier. Later, in seventeenth-century Lutheran orthodoxy, the distinction comes to be that between *providentia ordinata* and *providentia specialis*. See Bernhardt, *Was heißt 'Handeln Gottes'?*

[9] See Funkenstein, *Theology and the Scientific Imagination*, 124–52 (quote at 151). Compare also Harrison, 'Voluntarism and Early Modern Science', 8–14.

relationship between the *potentia Dei absoluta* and the *potentia Dei ordinata* were popular in scholastic theology but were regarded as specious by most Protestant theologians, who preferred to restrict their discussion to the world as they encountered it.

One theologian who chose to emphasize special providence over general providence is Jacob Andreae, Chancellor of the University of Tübingen from 1561 to 1590. In his preface to a series of sermons considering celestial conjunctions and particularly an eclipse due to take place in 1567, Andreae offers a swingeing critique of astrology, drawn largely from Pico della Mirandola's *Disputationes adversus astrologiam divinatricem*.[10] This critique is at the same time an attack on the usefulness of seeking to understand or interpret the natural world through observation, for Andreae, like Luther, does not believe that characteristics of God or messages for humankind can be gained from observations of the natural world. He justifies his view theologically by emphasizing God's ability to intervene in the 'normal' running of the world: since it is always possible that God might intervene in any causal process, it is ridiculous to think that any natural occurrence, such as a planetary conjunction, could have any meaning for one person or one particular group of people. Andreae is here criticizing the kind of providential astrology that had been defended by Philip Melanchthon, and which was appealed to by Andreae's colleague in Tübingen, Jakob Heerbrand,[11] and in doing so he emphasizes the unpredictability of the world and of God's providential action. The natural world, he believes, if it is to be observed at all, must be observed for its own sake and not because those observations will bring the observer closer to God. Since planetary conjunctions are predictable by observation of the order of the heavens, and thus in any case not believed to be the product of direct divine intervention, Andreae seems here to be suggesting that not even the study of the order of the universe can be judged to reveal the mind of God. Andreae's conviction that God can and does choose to intervene at any time in the normal order of the world, probably combined with the fundamental Lutheran conviction that the dichotomy between law and gospel must render the findings of philosophy illegitimate as theological arguments, leads him to reject any attempt to come closer to understanding God through the observation of the natural world.[12]

A generation later, the young theology professor Matthias Hafenreffer was to take a somewhat different position, although he too emphasized the role of *providentia specialis*. Although in his theology textbook Hafenreffer

---

[10] Andreae, *Christliche, notwendige vnd ernstliche Erinnerung*, Aa2^r–Eeii^r.

[11] Methuen, *Kepler's Tübingen*, 132–52.

[12] A more detailed consideration of these aspects of Andreae's theology is offered in Methuen, *Kepler's Tübingen*, 125–8.

pays lip service to the possibility that the *liber naturae* will offer a source of revelation about God, citing only scriptural evidence for its importance and no arguments from the order of the natural world itself,[13] he does assert elsewhere that divine providence may be seen in the regularity of the created universe, notably in the motions of the celestial bodies and in the order of society, all of which were created by God (and which are generally understood to be part of the *liber naturae*).[14] But at the same time, Hafenreffer, like Andreae (and perhaps in reaction to the work of his friend and former student Johannes Kepler), also emphasizes God's special providential action in the world. Hafenreffer is convinced that God may choose to act above the laws of nature, because God is not bound by secondary causes,[15] so that God may choose at any time to overturn the normal running of the world. It is, therefore, ultimately impossible for a human observer to grasp the order of nature through such study. In the theology of Andreae and Hafenreffer, then, the conviction of the ultimate futility of seeking to understand the workings of God's providence through a study of the natural world seems closely linked to their belief that God can and will interfere in the order of nature.

A somewhat different emphasis can be observed in the theology of Jakob Heerbrand, Andreae's successor as Chancellor of the University of Tübingen, student of Melanchthon and teacher of both Hafenreffer and Johannes Kepler. In his theology, Heerbrand places a strong emphasis on the book of nature, the *liber naturae*, as a divine source of revelation about God which is in some ways parallel to the Bible, the *liber scripturae*. For Heerbrand the *liber naturae* explicitly includes both the natural world and the order of society, but in contrast to Hafenreffer, Heerbrand really does seek to understand what the contents of this book might be.[16]

For Heerbrand, the praise of God is the aim and purpose of creation. But according to him, it is primarily the order which can be found in nature which demonstrates God's continuing care of the world and thus his providence. This *providentia Dei* can be seen from a number of sources: the order and beauty of the world, the various sorts of animals and plants, the ability of the human mind and human reason to know God and to distinguish between good and evil. The aspects of divine providence that are made visible through the order of creation demonstrate that a fundamental

---

[13] Hafenreffer, *Loci theologici*, esp. 2–3. This aspect of Hafenreffer's theology is considered in Methuen, *Kepler's Tübingen*, 116–17 and 151–2.

[14] Hafenreffer, *Disputatio de providentia Dei*, 2^r.

[15] Hafenreffer, *Loci theologici*, 17; cf. idem, *Disputatio de providentia Dei*, 5^r.

[16] Heerbrand, *Compendium theologiae*, but compare also his sermon on the comet of 1577/8, *Ein Predig von dem erschrockenlichen Wunderzeichen am Himel*, and his *Disputatio de Magia*. For Heerbrand's theology see Hübner, *Die Theologie Johannes Keplers*, 194–9, and Methuen, *Kepler's Tübingen*, 110–13, 137–52.

order underlies the natural world. But God also makes himself known in a different way, through miracles and direct interventions, which equally show his care for the world but which break through its customary order (e.g. the gift of manna from heaven to the people of Israel in the desert).[17] It seems to be an intervention of this type that he has in mind in his sermon about the comet of 1577/8. Heerbrand seems to envision the appearance of the comet as a divine decision to act directly and through a direct manipulation or application of natural causes.[18] Phenomena such as this comet thus appear in the natural world and can be understood as arising from natural causes, but are nonetheless the result of unpredictable, arbitrary interventions by God.

Heerbrand does not offer a broader philosophical basis for his theology. However, when compared to the theology of Philip Melanchthon, it can be seen that Heerbrand takes a similar approach to that of his teacher. Melanchthon understands the usual definition of providence to be 'the knowledge by which God discerns and provides for all things, and the government by which he maintains the whole of nature'.[19] Knowledge of the natural world, in particular that gained through observation of the human body or of the perfect, supra-lunar heavenly sphere, is an essential ingredient of recognizing these aspects of providence. Besides allowing the measurement of time, and thus enabling much of the rhythm of human life, the perfect, orderly movements of the heavenly bodies make visible the order that God wants for human, earthly society, and as such the natural world reflects the providence of God. The movements of the heavens should be observed with precision with the tools of philosophy and mathematics, the means by which the human mind is raised to God.[20] The orderly movements of the heavens are thus understood to reveal God and to incite us to precise observation and study. Melanchthon, who refers to Plato and Pythagoras, understands the structures of the universe to be mathematical in character, and as such not only to reflect the divine mind, but also to be identifiable by the human mind, created *imago Dei*.[21] In these arguments, Melanchthon, like Heerbrand, argues that it is the order of the universe which allows it to witness to divine providence.

That this order can be interrupted is clear to Melanchthon, but he

---

[17] Heerbrand, *Compendium theologiae*, 176–8.
[18] Heerbrand, *Ein Predig von dem erschrockenlichen Wunderzeichen am Himel*, 1–8.
[19] Melanchthon, *Initia doctrinae physicae*, CR 13, 203.
[20] Melanchthon, *Prefatio in arithmeticen*, CR 11, 288. Melanchthon's use of astronomy and of anatomy has been discussed in detail by Kusukawa, *The Transformation of Natural Philosophy*; for his astronomy, cf. also Methuen, *Kepler's Tübingen*, 61–106.
[21] For God as *mens*, see Bellucci, 'Gott als mens'. For the human ability to recognize the divine mind, see Bellucci, *Science de la Nature et Réformation*, esp. 321–480. Cf. also Frank, *Die theologische Philosophie Philipp Melanchthons*, 102–11.

believes that exceptions do not negate the underlying rule.[22] Moreover, he is also convinced that such exceptions do not come about by chance. God can intervene in an unexpected manner, for instance when he arranges the appearance of a comet. Such phenomena would have been unnecessary had it not been for the Fall; they are witnesses to the will of God and act as a warning against unchristian lifestyles or as an encouragement of the practice of true piety.[23] There seems to be an implication here, as there is for Heerbrand, that although phenomena such as comets exist and can be understood in the context of the natural order of causation (understood in terms of Aristotelian physics and theories of causality), they are not part of the orderly progression of the course of the world, because they arise unexpectedly. But the Christian astrological interpretation of such special phenomena as revealing the will of God, which for Melanchthon forms part of physics, is theologically possible precisely because he believes that these phenomena must be understood as a part of the workings of divine providence.

Thus, it is not only the order of the heavens and the design of the human body – that is, the orderly workings of the universe – which are important in discovering God's providence; unusual divine interventions are equally evidence of God's care for the world. Melanchthon's thought implies that the workings of special providence are necessary only because fallen human nature can have only an imperfect grasp of the true order of the world. Such interventions would have been unnecessary before the Fall, since human nature could have understood God's will from the normal running of the world. But because we are fallen, human observers who wish to understand the will of God must be attentive to both aspects, that is to the workings of both special and general providence.

Both Melanchthon and Heerbrand integrate their understanding of the importance of observing the heavens into a Christian astrology which allows a theoretical underpinning to their conviction. The theology of Heerbrand and of the Lutheran (albeit later regarded as crypto-Calvinist) Reformer Melanchthon thus offers a contrast to that of Andreae and Hafenreffer, which not only demonstrates the difficulty of codifying 'Lutheran' approaches to the natural world, but also indicates that those theologians who understood the natural world to be a part of divine revelation drew upon the conviction that all that happens in the universe can be regarded as revelatory, while those who were opposed to the idea of finding traces of God in the universe tended to emphasize God's power to intervene and thus to assume the futility of studying the normal order of the world. Heerbrand

---

[22] Frank, *Die theologische Philosophie Philipp Melanchthons*, 286.
[23] Methuen, *Kepler's Tübingen*, 85–8.

and Melanchthon are not solely concerned with the observation of the order which they perceive to be present in creation and in providence, for they believe that it is the creation as a whole, that is, both the normal order of the world and the interventions of God described by special providence, which must inform the observer's understanding of God. On the other hand, they also indicate that it is the order underlying general providence which demonstrates God's true intentions for the world. God's special interventions take place when things go wrong and are thus reminders and warnings for particular situations, rather than models for the ethical order of society or indications of the fundamental order in the mind of God (and thus in the creation). Such interventions have something to tell the observer about the will of God, but they can only be interpreted retrospectively and cannot be used for prediction.

'Providentia specialis' and the nova of 1572

These theological discussions contributed to the context in which sixteenth-century astronomers were working. While not every astronomer was also a theologian, some were.[24] Moreover, practically every astronomer was concerned not only to observe but also to interpret his observations either astrologically or theologically.[25] Sixteenth-century astronomers were concerned with both aspects of God's providence, for they sought to observe, to understand and to predict the normal course of the heavens, basing their observations and predictions mostly on the work and theories of Ptolemy, while at the same time they were also concerned to observe and to interpret phenomena such as comets and meteors, which, although generally understood in the terms of Aristotelian physics to be sublunar phenomena arising from natural causes (the ignition of gases), appeared in the skies and thus came under the remit of the astronomer. How, then, did such theological considerations affect the approach of astronomers?

As discussed in the previous chapter, one widely observed phenomenon which caused considerable problems of interpretation was the 'comet or new star' of 1572. Although many treated it as a sublunar comet, it was observed by a number of astronomers to be above the moon. As such it brought change into the perfect, supra-lunar realm, where (at least according to contemporary interpretations of Aristotle) no change was supposed to take place. As shown above this phenomenon was observed to

[24] Michael Maestlin and Johannes Kepler both studied theology at the University of Tübingen under Jacob Heerbrand.
[25] For these aspects of the role of sixteenth-century astronomers, see Schöner, *Mathematik und Astronomie an der Universität Ingolstadt*, esp. 96–103, and Westman, 'The Astronomer's Role'.

be supra-lunar by a number of astronomers and mathematicians, among others Philip Apian and Michael Maestlin in Tübingen, Caspar Peucer in Wittenberg, and Wilhelm of Hesse in Kassel.[26]

In their observation and subsequent discussion of this phenomenon, all four of these astronomers conclude that they are observing a phenomenon which occurs beyond the moon[27] and thus represents a change in the supra-lunar region. With the exception of Maestlin, they also understand the phenomenon as a warning from God. Thus Apian explains that 'before he sends sickness and punishment, the merciful Father usually sends a good preacher to the pious and to the Godless, which is intended to convince and comfort the pious in the fear of God',[28] while Wilhelm of Hesse regards it as an omen of the approaching apocalypse and immediately writes to his cousin, admonishing him to prepare himself for the Last Judgement.[29] Both astronomers thus understand this phenomenon to be the result of a direct intervention by God.

The conclusion that the phenomenon is beyond the moon poses a problem to Wilhelm of Hesse and to Peucer, whose understanding of Aristotelian natural philosophy excludes the possibility that change can take place outside the sublunar sphere. Their solution to this problem is to turn to the doctrine of special providence, which allows them to accept the theoretically impossible conclusions drawn from their observations, while leaving intact the theoretical system of physics, which could thus continue to guide their interpretation of natural observations. Recourse to special providence could therefore serve to legitimate observations which would otherwise cause a system to break down.

Clearly, if the fundamental system of interpretation is adjusted to legitimate such observations, they no longer present a problem. This is recognized by Maestlin, who in his published consideration of the 1572 nova also notes that according to the theories of Aristotle and Ptolemy, the appearance of a new star (for he is sure that that is what it is) is impossible, so that it cannot be understood to be the result of natural causes as traditionally understood.[30] However, although Maestlin does understand such phenomena to be manifestations of God's will, he does not appeal to the intervention of God to legitimate his observations. Instead, he seeks to

---

[26] These astronomers conducted a correspondence with Ludwig, Duke of Württemberg, discussed in detail in Chapter 4 above.

[27] Apian and Maestlin place the phenomenon in the sphere of the fixed stars, although Apian inclines to the view that it is a *cometa secunda*. Peucer and Wilhelm, who refer to the phenomenon as a *stella nova*, in fact place it in the sphere of Venus. See Chapter 4 above.

[28] Apian to Ludwig, 26 December 1572, Hauptstaatsarchiv, Stuttgart (HStAS), A274, Bü 21, fo. 5^r–^{v}.

[29] Wilhelm to Ludwig, 14 January 1573, HStAS, A274 Bü 21, fo. 1^r.

[30] Maestlin, *Demonstratio astronomica*, 28–9.

modify his system: he notes that Copernicus teaches that change may be possible in the supra-lunar sphere and concludes that there must be 'hyperphysical' causes which explain the appearance of this phenomenon, although he himself does not know what they are.[31] For Maestlin, the kind of appeal to special providence which asserts that a phenomenon has no natural causes is inadequate.

In a similar way, Philip Apian, who rejects both the Aristotelian theory that all comets must be sublunar and his father's suggestion that every comet must have a tail,[32] has no need to legitimate his observations through special providence. Although he considers it impossible to interpret the phenomenon properly during its appearance, since it can only be accurately placed in the context of eclipses and conjunctions after it has disappeared from the heavens,[33] Apian yet seeks to understand the phenomenon in terms of the normal course of the heavens, noting that it was preceded, as he believes a comet's appearance often is, by a major conjunction of the planets Jupiter and Saturn. Moreover, the winter was very cold, and Apian believes that such weather often results from the appearance of a comet.[34] Because Apian has already modified his understanding of the physics of comets to extend beyond the sublunar sphere, he needs to invoke special providence only when explaining why the nova appears at this particular time.

It can thus be seen that from the point of view of observational astronomers, the events of special providence may fall into two categories. More frequent are the events that can be explained in terms of natural causes, but which cannot be predicted. The other type is represented by the kind of event that completely defies explanation through natural causes, notably within the system of Aristotelian physics. Events of the latter category can be moved to the former category once it is recognized that the fundamental system of explanation must be modified. But as long as the explanation of special providence is taken to be valid, there will be no modification in the underlying interpretative system even when observations contradict it. Thus, if accurate observations are to lead to the modification of a previously accepted system of explanation, the making of observations that defy the system must be coupled with an expectation that events should in general be able to be assigned natural causes. This in turn tends to shift the observed phenomenon out of the area of special providence into

---

[31] Maestlin, *Demonstratio astronomica*, 30.

[32] Apian to Ludwig, 26 December 1572, fos 2^v–3^r; and compare the version published by Brahe in *Astronomiae Instauratae Progymnasmatum, Opera Omnia*, 3, 159–60.

[33] Apian to Ludwig, 26 December 1572, fo. 4^{r-v}.

[34] Apian to Ludwig, 26 December 1572, fos 1^v–2^v; cf. Brahe, *Astronomiae Instauratae Progymnasmatum, Opera Omnia* 3, 158–9.

the realm of general providence. God's special intervention is no longer required.

A similar process can be observed in Maestlin's discussions of the comets of 1577/8 and 1580. Although Maestlin regards the comet as a warning from God, he also seeks to define an orbit for the comet.[35] The attempt not only to observe the path of the comet but to define an orbit for it brings it closer to the status of the planets and other parts of the ordered, predictable movement of the heavens. Of course, if a closed orbit could be found it would be possible, at least in theory, to predict the recurrence of the comet (a possibility that did not occur to Maestlin). Intentionally or not, his search for the comets' orbits is thus effectively a search for a means of making such phenomena predictable. This is the kind of process which could move celestial events from the realm of special providence into the realm of mathematical representation and as such allow them to be understood in terms of general providence. It is surely no coincidence that Maestlin's theological justification for what he is doing bases itself upon Old Testament wisdom theology, with its strong conviction that God formed the order of creation, and that he consequently emphasizes the order of the course of the universe rather than the possibility of God's intervention.

Conclusion

The belief that the natural world was created by God leads to interpreting as revelatory, not just its order and regularity, but also the events of the *providentia specialis*. There is, however, a theological overtone that God's interventions, that is the events witnessing to the *providentia specialis*, are necessary mainly because God must intervene to re-establish what has been corrupted by the Fall. But it is the order of the *providentia generalis* which indicates God's original creative plan. Theologically speaking, the predictable regularity and order of the natural world witnesses to *providentia generalis* rather than to *providentia specialis*, and theologians who emphasize the latter over the former tend also to reject the study of the natural world as irrelevant.

The consideration of interpretations offered of the nova of 1572 shows that the responses to what was an inexplicable phenomenon in terms of Aristotelian physics depends upon the relative weight given to the explanations of physics or astronomy as opposed to those of theology. If

---

[35] See Maestlin, *Observatio et demonstratio cometae* and *consideratio et observatio cometae*. This endeavour is described in some detail by Westman, 'Johannes Kepler's Adoption of the Copernican Hypothesis', 39–66.

Aristotelian physics is accepted, the arguments of special providence may be used to legitimate an observation or a conclusion which would otherwise be excluded by the predominant theories of physics. Alternatively, if the underlying physics has already been revised, special providence may be used to explain why a celestial event occurs at this particular time. Or again, the inexplicable phenomenon may be regarded as evidence which requires the revision or modification of certain assumptions or theories, which can in turn lead to a revision of what may be understood as inexplicable.[36] At the same time, the appeal to arguments from physics correlates with certain theological convictions. If God is understood as capable of intervening arbitrarily at any time, such inexplicable phenomena present little challenge to the theoretical system. If, on the other hand, God is understood as acting predominantly within and through established and orderly systems accessible to the human mind, then unpredictability or unexpected observations become problematic. If God is regarded as generally working within the framework of natural causes, but as choosing to intervene every now and then in a totally radical way, then room is left for the acceptance of the validity of observations which would otherwise have to be rejected on the basis of physics.

Because most of the phenomena of *providentia specialis* could in fact be understood in terms of Aristotelian physics, the distinction between *providentia generalis* and *providentia specialis* is not the same as that between the explicable and the inexplicable. Nevertheless, some astronomers did appeal to the doctrine of *providentia specialis* to justify observations that would otherwise have been regarded as illegitimate on physical grounds. This allowed discrepancies to stand without contradicting the interpretative system, and as such the doctrine of special providence could allow – and indeed encourage – observations which were not bound by one particular interpretative system. It can therefore be seen that in a context in which the doctrine of special providence is taken seriously, accurate observations which may conflict with an accepted system of interpretation may be accepted as correct without necessarily calling the system into question.

Of course such observations may put so much pressure upon the underlying system of interpretation that observers begin to question it, as can be seen in the cases of Maestlin and Apian. The observation of the events of special providence may, therefore, bring about an adjustment which requires a shift in the understanding of certain phenomena which will move them from the sphere of special providence to that of general

---

[36] This range of reactions is typical for observers of phenomena which are problematic within an accepted scientific system. See, for instance, Kuhn, *The Structure of Scientific Revolutions*, 52–91.

providence. The requirement for this shift is the adjustment of the underlying interpretative system, which in turn requires the understanding that such a system should offer an order which admits as few exceptions as possible. And this presupposes an understanding that God will not intervene in that underlying order, even though that order may itself be understood as a witness to God.

6

Time Human or Time Divine: Theological Aspects in Opposing the Gregorian Calendar Reform

On 24 February 1582, Pope Gregory XIII issued a bull, *inter gravissimas*, initiating the reform of the Julian calendar. The bull laid down that in order to bring the calendar into alignment with the astronomical year, 4 October should be immediately followed by 15 October; in order to maintain that correlation three leap years were to be deducted every four hundred years; and in order to improve the correlation between the solar and the lunar calendars, a revised system of 'epacts' was to be introduced. The new Gregorian calendar was adopted by most (but not all) Catholic states,[1] and was rejected by most (but again not all) Protestant states.[2]

Gregory XIII's bull was issued at a time when the theological status of observations of the heavens was a controversial issue amongst Lutheran theologians. This was part of a quest, shared by theologians, philosophers and astronomers, for an authority – and a truth – which could be regarded as certain. The Reformation had set biblical revelation against the authority of the Pope and of councils and was understood to form the measure against which all other truth should be judged. But if theologians found

---

[1] The new calendar was immediately adopted in such predominantly Catholic states as France, Italy, Luxembourg, Portugal and Spain. Belgium, the German Catholic states and the Catholic part of the Netherlands followed suit in 1584. Hungary did not move to the new calendar until 1587, and in the city states of Florence and Pisa it was not adopted until around 1750.

[2] For instance, Augsburg, an imperial city which at that time was predominantly Protestant, adopted the calendar. However, most Protestant countries, including the German Protestant states, did not make the change until the turn of the seventeenth century, and Britain did not adopt the new calendar until 1752.

their highest authority in the Bible,[3] philosophers and astronomers found themselves seeking a philosophy and a method of deduction which would allow them to move away from what were often termed the 'sophistic arguments' of scholastic philosophy. Increasingly this had the effect that observational science began to change its status and to gain in authority. Movement towards the establishment of observational accuracy as the primary criterion for judging between systems can already be observed in discussions of the nova of 1572, or the comet of 1577/8, but this was a slow process which involved not only the overturning of accepted physical theories, but the establishment of new priorities for the criteria by which theories should be judged, including that of observational accuracy.[4]

In general, however, although the practical application of these principles might – and did – alter understandings of cosmology, it did not introduce any immediate or radical change into daily life. The reform of the calendar was, of course, very different, since it demanded an immediate alteration of the way that the year was to be reckoned. This controversy thus offers the scholar a chance of assessing the status of arguments of astronomical accuracy compared to other criteria. This chapter considers the arguments used against adopting the calendar reforms, focusing on treatises written by two men deeply interested in the revelatory potential of the natural world[5] – the theologian Jacob Heerbrand, Professor of Theology at the University of Tübingen,[6] and the mathematician and astronomer Michael Maestlin, trained at Tübingen but at that time Professor for Mathematics at the University of Heidelberg.[7] In particular this chapter considers the way in which theological arguments were applied to this question – not, at first sight, a theological issue at all. It begins, however, with a brief excursion through the developments leading to the proclamation of Gregory XIII's bull.

The dating of Easter

The central reason for the calendar's importance to the Church was the necessity – and complexity – of determining the date of Easter. As laid down by the first Council of Nicaea (325 CE), Easter Sunday falls on the Sunday immediately following the full moon on or after the vernal equi-

---

[3] See, for instance, Heerbrand, *De scripturae sacrae interpretatione*, 5, thesis 35.
[4] Methuen, *Kepler's Tübingen*, esp. 159–204.
[5] See Methuen, *Kepler's Tübingen*, esp. 107–58.
[6] Heerbrand, *De adiaphoris*.
[7] Maestlin, *Ausführlicher und Grundtlicher Bericht*; idem, *Alterum Examen novi Pontiocialis Gregoriani Kalendarii*.

nox, defined by the Council to be 21 March.[8] The accurate determination of the date of Easter thus presupposes, firstly, a means of correlating the solar and lunar calendars and, secondly, a calendar which calculates the year so as to place the vernal equinox reliably on 21 March.[9] Neither of these presuppositions is trivial. The Julian calendar, which formed the basis for the calculations of Easter,[10] includes a leap year every fourth year and thus assumes an astronomical or tropical year of exactly 365¼ days. However, since the length of the tropical year is (by current reckoning) 365.242 days,[11] the Julian calendar introduced an error of 11 minutes and 14 seconds each year, resulting in a cumulative error of almost one and a half days in two centuries, and seven days in a millennium. Thus, by the mid-sixteenth century the calendar was placing 21 March ten days later than the actual occurrence of the vernal equinox. Equally problematic was the correlation of the solar and lunar years. During the Middle Ages, the date of the full Easter moon was calculated by means of 'golden numbers' introduced in 530 CE. However, these were inaccurate and yielded an error of up to two days in the prediction of the lunar cycle.[12] The combination of the two errors made determining the date of any given full or new moon very difficult, and the date fixed for Easter became more and more arbitrary, and less and less related to the occurrence of the full moon.

Although the discrepancies between prediction and reality in the lunar cycle had been noted in the eighth century by both Bede and Alcuin,[13] it was not until the thirteenth century that the problems inherent in the Julian calendar began to become apparent.[14] However, contemporary observational methods made it very difficult to calculate either error accurately, so it was not possible to find a sustainable correction. Nevertheless, by the

[8] This definition was accepted by the Western churches; the Eastern churches (and until the sixth century also the church in Britain) adopted a calculation of Easter based upon a different understanding of the counting of days in the Jewish calendar.

[9] The vernal equinox is the moment at which the centre of the sun appears to cross the Equator as it moves from south to north.

[10] For the different calendars and methods of numbering years in use during the early Christian period, see Meyer, 'Chronologie IV', in *Die Religion in Geschichkte und Gegenwart*, 3rd edn (Tübingen: Mohr Siebeck, 1957–65), vol. 1, pp. 1814–17.

[11] The tropical year is defined as the length of time between two successive vernal equinoxes. Since this period fluctuates from year to year, it is calculated as the mean taken over several years. The length of the tropical year is, however, decreasing slowly. The tropical year in 1582 was probably 365.24222 days, whereas in 2000 it was 365.24219 days.

[12] The problem is complicated by the fact that the lunar cycle varies depending on the latitude of observation.

[13] Kaltenbrunner, 'Die Vorgeschichte', 291–2.

[14] Kaltenbrunner, 'Die Vorgeschichte', 292–315. Drawing upon knowledge of Arabic astronomy, a number of astronomers and natural philosophers suggested corrections to the calendar, including John Sacro Bosco (297–9), Johannes Campanus (300–3), Robert Grosseteste (304–7) and Roger Bacon (310–14). For summary of the arguments presented from the time of Grosseteste, compare also North, 'The Western Calendar'.

fourteenth century, the error in the prediction of the new moon had led a number of mathematicians to devise calendars for local use; these differed, sometimes radically, from the ecclesiastical calendar to be found in missals and breviaries.[15] For this reason (and presumably because of the problems inherent in using a number of calendars alongside one another), the inaccuracies of the calendar began to concern the ecclesiastical hierarchy. In 1345 Pope Clement VI of Avignon commissioned a treatise on the subject.[16] At the suggestion of Pierre d'Ailly, a decree for calendar reform was issued by Pope John XXIII in 1412; this was debated at the Council of Constance (1414–18), but John's problematic status as one of the schismatic popes meant that it was not enacted.[17] At the Council of Basle (1431–49), a petition was brought that the calendar be reformed, and the proposals of Nicolas of Cusa and Hermann Zoestius were debated.[18] However, there was still no agreement about the size of the error, and no changes were implemented.[19] In 1476, Pope Sixtus IV called the mathematician Johannes Regiomontanus to Rome to advise him on calendar reform, but Regiomontanus died soon after reaching Rome,[20] and the matter was once again allowed to lapse.

Under pressure from Paul of Middelburg, the question arose again at the Fifth Lateran Council (1512–17) under Pope Leo X.[21] At Leo's request, Emperor Maximillian I commissioned reports on the calendar from the universities of Vienna, Tübingen and Leuven.[22] At the council, however, the calendar debate was postponed, initially to another session and then indefinitely. It was not until the final session on the final day of the Council of Trent, 4 December 1563, under the leadership of Pope Pius IV, that it was resolved that the breviary and the missal should be reformed, a decision which implied that attention must be paid to the calendars upon which they were based.[23] However, decisive action had to wait until 1582, the year of accession of Pope Gregory XIII, who commissioned the Jesuit mathematician Christopher

---

[15] Kaltenbrunner, 'Die Vorgeschichte', 355, 357, 374.
[16] Kaltenbrunner, 'Die Vorgeschichte', 315–22.
[17] For the work of Pierre d'Ailly, see Kaltenbrunner, 'Die Vorgeschichte', 326–36.
[18] Kaltenbrunner, 'Die Vorgeschichte', 336–54.
[19] Cusa's suggestion that ten days be omitted from the calendar in order to ensure that the vernal equinox should fall on 21 March as envisaged by the Council of Nicaea was eventually adopted, but at this stage was still regarded as unnecessary.
[20] For Regiomontanus' work and his calendar, see Kaltenbrunner, 'Die Vorgeschichte', 367–74.
[21] For the work of Paul of Middelburg, see Kaltenbrunner, 'Die Vorgeschichte', 375–85. The council had been summoned by Pope Julius II, who also succeeded in winning the support of the emperor for this endeavour.
[22] Reports were written by Georg Tannstedter and other students of Regiomontanus – to which Georg Peurbach in Vienna responded (Kaltenbrunner, 'Die Vorgeschichte', 386–90) – and by Johannes Stöffler in Tübingen (390–5). Leo X wrote to other universities asking their opinion, but no replies are extant.
[23] North, 'The Western Calendar', 101.

Clavius to draw up a satisfactory correction based upon the observations of the astronomer-physician Luigi (or Aloysius) Lilius of Verona (1510–76).[24]

Luther and the calendar

Throughout the centuries they were concerned with the question, the popes had been able to assume that the calendar might be seen as a matter for regulation by the Church. However, in the course of the sixteenth century, the Western church was ruptured by the Reformation, and Protestants began to question the authority of either the Pope or an ecclesiastical council to act on this matter. As early as 1539, Luther had included in his treatise *Von den Konziliis und Kirchen* a discussion of the problems of the calendar.[25] Focusing on the date of Easter as defined by the Council of Nicaea, Luther argued that the council had sought to use new cloth to patch old:

> They want to retain a part of the old law of Moses, namely to pay heed to the full moon in March – that is the old garment then (as Christians delivered by Christ from the law of Moses by Christ), they do not wish to be subject to that same day of full moon, but instead want to take the following Sunday – that is the new patch on the old garment.[26]

It would be much better, Luther suggested, to drop the 'law of Moses' from this discussion entirely, and to celebrate Easter on a fixed day of the year rather than on a particular Sunday, just as Christmas is always celebrated on 25 December:

> We therefore have and must have the power and the freedom to observe Easter when we choose, . . . Moses is dead and buried through Christ, and neither days nor seasons should be lords over Christians, but rather Christians are lords over the days and seasons, free to fix them as they will or as seems convenient to them. For we know we shall attain salvation even without Easter and Pentecost, without Sunday and Friday, and, as St Paul teaches us, we shall not be damned on account of Easter or Pentecost, Sunday or Friday.[27]

While Luther accepted the observations of the astronomers that the calendar was no longer in step with the astronomical year, he commented that the slippage in the calendar meant that all saints' days and festivals,

---

[24] For the role of Clavius in the reform, see, for instance, Baldini, 'Christoph Clavius'. For Lilius, see Moyer, 'Aloisius Lilius'.
[25] Luther, *Von den Konziliis und Kirchen*, WA 50, 554–9; LW 41, 61–8.
[26] Luther, *Von den Konziliis und Kirchen*, WA 50, 555–6; LW 41, 63.
[27] Luther, *Von den Konziliis und Kirchen*, WA 50, 558–9; LW 41, 67.

and not only Easter, were being celebrated too late. Simply to disregard this slippage, or to celebrate Easter on a set date, would be to underline the fact that every day, and not only Sunday, Easter or particular festivals, must be understood – and lived – as the Lord's Day.[28] Moreover, his expectation that the end of the world was approaching inclined him to make light of the problems: 'the old garment with its great rents has stayed on and on and now it may as well stay until the Last Day, which is imminent'.[29]

For Luther the discussion of the date of Easter and the marking of particular days raised important questions about the relationship of the old (Mosaic) law to the new, about the status of the keeping of such feasts, and about the importance of living every day as the Lord's Day. These were all issues which would not be settled by the introduction of calendar reform, and in some cases a new calendar with a consciously more accurate link to the astronomical year, and thus the ability better to represent certain days, could even be seen as a retrograde step. Luther was writing over forty years before the Pope issued the bull announcing the calendar reform, at a time when no solution had been offered, but it is clear that he would not have been an unconditional supporter of the reform to the calendar.

Protestant responses to the Gregorian reform

When it came, the Pope's reform was greeted by a spate of Protestant responses.[30] Amongst these was a reprint of the relevant extract from Luther's *Von den Konziliis und Kirchen*, with the addition of a paragraph which raised the question which had now become crucial to the Protestant camp:

> One does not ask whether it is right or not right to reform and alter the calendar as it has been used until now, but who should do it, such as their majesties king and emperor. The pope is not a king and not an emperor, although he would like to be. Therefore it behoves him not.[31]

The second question raised by the interpolation was that of how a change in

---

[28] Luther, *Von den Konziliis und Kirchen*, WA 50, 557–9; LW 41, 61–7.
[29] Luther, *Von den Konziliis und Kirchen*, WA 50, 557; LW 41, 65.
[30] Thirty-two treatises discussing the calendar reform and published by Protestants between 1583 and 1596 are listed by Stieve, *Der Kalenderstreit*, 89–98. To Stieve's list must be added Maestlin's *Alterum Examen novi Pontiocialis Gregoriani Kalendarii* (1583). Joseph Scaliger's chronological comparison of the histories of different peoples, *De emendatione temporum* (1583, which went into several editions), also included a discussion of the calendar reform.
[31] Anonymous, *Herrlich Bedencken*, quote from fo. B^v: 'Denn man fraget nicht / obs recht oder vnrecht sey / das man den bishero breuchlichen Calender / Reformire vnd endere / sondern Wer es thun sol / Als Nemlich / die Mayesteten Keyser / Könige / Der Bapst ist kein König / kein Keyser / ob er es gleich gern sein wolte / Derwegen gebüret es im nicht'.

the calendar should be introduced, if it were to be decided that such a change was necessary. Changes to the calendar should not be introduced piecemeal, concludes 'Pseudo-Luther', for then chaos would ensue.³²

The interpolation, although it did not originate from Luther's pen, was surely in the spirit of Luther. Ferdinand Stieve has suggested that the 'true reason' for Protestant questioning of the legitimacy of the Pope's command that the calendar be reformed was their hatred of the Pope and the suspicion that the calendar reform was a papal plot to reimpose his authority over Protestant churches and states.³³ Certainly, Heerbrand's suspicion of the Pope permeates his arguments against calendar reform. Heerbrand prefaces his disputation with the warning that the new calendar is the work of 'the anti-Christ and the devil';³⁴ he contends that by believing himself able to 'change the seasons', the Pope is usurping divine powers and so he lays himself open to the charge of being an anti-Christ.³⁵ However, this is not Heerbrand's main argument; taken as a whole, the disputation presents arguments which focus on the necessity of determining whether the calendar is an ecclesiastical or a civic matter. Ultimately, Heerbrand's reservations are rooted not solely in his suspicion of the Pope, but in the distinctions between Church and state which underlie Luther's theology of the two regiments.

Heerbrand sees problems in introducing calendar reform whether the calendar is regarded as political or as ecclesiastical. He is concerned that because it has been put forward by Pope Gregory XIII, and supported by the Council of Trent, the acceptance of the Gregorian calendar in Protestant states would be tantamount to acknowledging the authority of the Pope over either Germany or the Church.³⁶ If the calendar is a political matter, the authority responsible for introducing change should be the Emperor, and not the Pope. Moreover, although Heerbrand recognizes that the calendar is in many ways a political measure, he protests that its association with the name of a pope, taken together with the arguments which have been brought to bear about the necessity of celebrating festivals and holy days at the correct times, and the accompanying threat of excommunication for those who do not obey the Pope's command,³⁷ show that it is as much ecclesiastical

---

³² Anonymous, *Herrlich Bedencken*, fo. Biiʳ: 'Wie man es thun sol: Eintrechtiglich, / das ein Gebot zugleich / in alle Welt ausgienge / von denen / so es befehl haben / Nicht das es hie in einem winckel der wolt / dort auch an einem'.

³³ Stieve, *Der Kalenderstreit*, 51.

³⁴ Heerbrand, *De adiaphoris*, Praefatio, 2–7.

³⁵ Heerbrand, *De adiaphoris*, 26, thesis 64. Heerbrand cites Daniel 7.25b in support of this claim. The polemical aspects of Heerbrand's treatise form the focus of Hoskin's discussion of the reception of the Gregorian calendar in Protestant lands ('The Reception of the Calendar by Other Churches', 259–61). However, Heerbrand's arguments are more varied than Hoskin's focus on his polemic would suggest.

³⁶ Heerbrand, *De adiaphoris*, 15, thesis 27.

³⁷ Heerbrand, *De adiaphoris*, 30–2, theses 87–99; 41–2, theses 143–6.

as political. This being the case, the Pope is still seeking to make decisions in an area which is not his to decide: the introduction of the calendar can be compared to the opposition which the churches of the *Confessio Augustana* have experienced as they seek to introduce changes to church practice.[38]

But nor is the calendar a matter of evangelical faith. Following Luther, Heerbrand argues that the dates of festivals cannot be regarded as essential to the gospel, since the Church has always met at different times and in different places as circumstances permitted. Thus the early church met in secrecy at night in private houses or in underground crypts, but this would no longer be seen as appropriate in the sixteenth century.[39] Equally, the churches of the East and West have celebrated Easter on different days for centuries; although this is undesirable, it has no dire effects, for what is important is not that there should be similarity of ceremonies throughout the Church, but that the doctrine should be one.[40] To support this opinion, Heerbrand argues that matters pertaining to the calendar may be regarded as belonging to the adiaphora, that is, to 'those things which the Word of God neither prohibits nor commands, and which can be either observed or neglected without sin', but which ensure that the Church is run with order and decency and to the edification of those who participate in its ceremonies.[41] The adiaphora, or 'things indifferent', were matters (generally church practices) about which concessions might be made by Protestants in the interests of peace without endangering Protestant doctrine. By accepting the calendar as adiaphora, Heerbrand in effect accepts that if the calendar is to be regarded as (even in part) an ecclesiastical matter, it is quite legitimate for the Pope to make one decision whilst Protestant princes make another.

If, on the other hand, the calendar is a purely political measure, the Emperor must be involved in any reforms. Moreover, 'learned people teach that nothing in the political sphere should be subject to change unless there appears to be evidence for the utility of that change, or it seems that evil will of necessity result from preserving the status quo'.[42] In this case, Heerbrand asserts that since 'nothing of detriment has followed from the use of the ancient [Julian] calendar for so many centuries, whether in the Church or in

---

[38] Heerbrand, *De adiaphoris*, 38, theses 125–6.
[39] Heerbrand, *De adiaphoris*, 20, thesis 49.
[40] Heerbrand, *De adiaphoris*, 15, thesis 27.
[41] Heerbrand, *De adiaphoris*, 10, thesis 1; compare also 19, thesis 45. The *Formula concordiae* defines adiaphora as ceremonies of ecclesiastical practices 'which are neither commanded nor forbidden in the Word of God, but which are ordered only for the sake of goodwill and good order' (article X). Debates about the status of the adiaphora or *Mitteldinge* had grown in intensity with the reintroduction of certain Catholic practices during the Interim of 1548. The *Formula concordiae* concluded that in times of persecution no compromise should be made, but that at other times, local communities could make local decisions to build up the people of God (article X, Affirmativa, 2, 4).
[42] Heerbrand, *De adiaphoris*, 51, thesis 184.

the republic or in the economic sphere, . . . there is no reason to change it on account of errors or rather, as Luther puts it, astronomical solecisms'; instead the old calendar should be retained until such a time as imperial authority should impose another alternative.[43] In any case, Heerbrand is not convinced that a true alternative has really been found, arguing that for every mathematician who appeals for the introduction of the new calendar on the basis that it will prevent further slippage of the year against the solstices and equinoxes, there is another who argues that it is impossible to introduce any correction which will enable the calendar to be perfectly correlated with the solar year.[44] In short, since it cannot be shown that the old calendar causes real problems or that the new one will bring any particular benefits, and since such a change is bound to bring with it terrible confusion, the calendar should be left as it is.[45]

It must be said that Heerbrand's disputation demonstrates neither a deep knowledge of nor real interest in the astronomical significance of the calendar: as we have seen, at one point he makes the mistake of seeing the reform of the calendar as an attempt to 'change the seasons'.[46] Instead, his interest in the calendar is primarily theological and political. In his attribution of the calendar to the adiaphora, Heerbrand takes up Luther's argument that celebrating particular feasts on particular days is not an essential part of the practice of faith, and places it in the context of current theological debates about the central tenets of faith and practice. Heerbrand uses this argument to place the calendar outside the Pope's jurisdiction regardless of whether it is a civic or an ecclesiastical matter, which enables him to counter the arguments on the basis of festivals and feast days used by the supporters of the change.

Michael Maestlin and the calendar

In contrast to Heerbrand, the astronomer Maestlin demonstrates a precise and historical understanding of the development of the calendar and the astronomical issues involved. In his treatises, he discusses in detail the development of the solar calendar and the reasons for introducing the leap year. Maestlin is fully aware that the calendar is a matter of convenience, set up to count the periods between particular solstices and other solar events. He knows that the current method of measuring the year gives rise to an error of one day approximately every 138 years, and he argues that, had

[43] Heerbrand, *De adiaphoris*, 51, thesis 183, and compare 32–3, thesis 100.
[44] Heerbrand, *De adiaphoris*, 55, thesis 194.
[45] Heerbrand, *De adiaphoris*, 51–3, theses 184–8.
[46] Heerbrand, *De adiaphoris*, 26, thesis 64.

Julius Caesar had access to the present state of knowledge at the time he introduced the calendar, it would have made perfect sense for him to introduce the refined, Gregorian system. Maestlin clearly understands the necessity of correlating astronomical observations with measurements of, in this case, the year. He offers a detailed discussion of the history of calendars, and in particular of the development of the first solar calendar by the Egyptians, who counted 365 days, and of the correction by the addition of a leap year every four years in the Greek calendar,[47] and he describes the differences between the lunar and solar year. Solar calendars such as the Julian calendar are characterized by the fact that the equinoxes (the days on which day and night are the same length) and the solstices (the days that are shortest or longest) recur on the same days every year, so that the seasons recur in the same months.[48] However, despite the leap year, it can now be observed that 'the sun in our times comes 12 or 13 days earlier'.[49] That is, the correction made by the Julian calendar does not suffice, because the year is in fact not 365 days and six hours long, but 365 days, 5 hours, 49 minutes, 15 seconds and 46 thirds long,[50] so that in a period of 134 years, one leap year too many occurs.[51] Moreover, there is also a mistake in the calculation of the golden numbers (which allow the correlation of the solar and lunar years), which means that the date of Easter is no longer being correctly calculated in the Western church.[52]

Maestlin's historical analysis demonstrates his understanding of the arbitrary nature of the choice to base the calendar upon the solar year and of when to start the new year.[53] Moreover, his detailed criticisms of the Julian calendar show him to be fully aware of the benefits in accuracy to be gained by adopting the Gregorian calendar. On the basis of this analysis of the shortcomings of the Julian calendar, it might be expected that Maestlin would press for the adoption of the Gregorian calendar. However, he argues against its adoption, asserting that although questions about the accuracy of the calendar have been raised for over two hundred years, every mathematician must be ashamed to press for the introduction of a new calendar 'in these blessed time'.[54] One of the reasons for this is that the calendar is being introduced by the Pope, who, in Maestlin's opinion, does not have the authority to determine how the movements of the sun and moon should

[47] Maestlin, *Ausführlicher und Grundtlicher Bericht*, 8–9.
[48] Maestlin, *Ausführlicher und Grundtlicher Bericht*, 11–12.
[49] Maestlin, *Ausführlicher und Grundtlicher Bericht*, 18.
[50] In the sexagesimal system, a third is one-sixtieth of a second.
[51] Maestlin, *Ausführlicher und Grundtlicher Bericht*, 19–20.
[52] Maestlin, *Ausführlicher und Grundtlicher Bericht*, 20–8.
[53] Maestlin appears to take the beginning of the calendar year to be 1 January, although in many areas of sixteenth-century Western Europe the calendar year was still deemed to begin on Lady Day, 25 March.
[54] Maestlin, *Ausführlicher und Grundtlicher Bericht*, fo.)()(i^{v}–)()(ii^{r}.

be postulated in the calendar.[55] As a Lutheran, Maestlin asserts that the Pope 'has removed Holy Scripture and the word of God from the sight of the common people and even forbidden them to read it', thus denying them the light of the gospel and knowledge of true salvation.[56] However, now that 'the Bible has come back into our hands', it is no longer necessary for faith to be passed down by the Pope, and he is no longer to be viewed with respect and awe.[57] Like Heerbrand, Maestlin is concerned that the Pope should not be allowed to believe that he can claim political authority in Germany.[58] In attempting to introduce a calendar reform, the Pope is usurping an authority which is not his to claim and which has been specifically denied by the Reformation. The acceptance of the new calendar would be to accord authority to the Pope and to his council, an authority which Maestlin and others in the Lutheran church are not prepared to give.

This is Maestlin's primary argument, but he can also enumerate a number of considerations which support his opinion that the Gregorian calendar should not be substituted for the Julian. These fall into three categories: political, ecclesiastical and mathematical.[59] Politically speaking, it is apparent that only scholars and educated people are aware of the problems with the calendar, for the shift of one day every 134 years is far too little to be noticed in a lifetime.[60] There is therefore little real need for a correction. Moreover, like Luther, Maestlin confidently expects the day of the Last Judgement to arrive soon, and at the latest in the year 2000. In consequence, a maximum of three additional days' error will accumulate, and for this it is certainly not worth introducing a correction.[61] The common people have no understanding of the golden numbers and are not in the least interested in the correlation between solar and lunar years. This too means that a change is unnecessary.[62] Economically, the introduction of a new calendar would bring chaos;[63] the farmers would not be able to pay their workers or the interest to their landlords, for the correction would mean that the days for payment fell before the harvest;[64] introducing such a correction would thus be more trouble than it is worth.[65] These are the worldly or political reasons why the calendar should not be changed.

Ecclesiastically, Maestlin, like Heerbrand – and like Luther before

---

[55] Maestlin, *Alterum Examen*, 50.
[56] Maestlin, *Ausführlicher und Grundtlicher Bericht*, fo.)(ii^{r-v}.
[57] Maestlin, *Ausführlicher und Grundtlicher Bericht*, fo.)(ii^v–)(iii^r.
[58] Maestlin, *Ausführlicher und Grundtlicher Bericht*, 102–8.
[59] Maestlin, *Ausführlicher und Grundtlicher Bericht*, 31–2.
[60] Maestlin, *Ausführlicher und Grundtlicher Bericht*, 36–7.
[61] Maestlin, *Ausführlicher und Grundtlicher Bericht*, 37–9.
[62] Maestlin, *Ausführlicher und Grundtlicher Bericht*, 39–40.
[63] Maestlin, *Ausführlicher und Grundtlicher Bericht*, 41.
[64] Maestlin, *Ausführlicher und Grundtlicher Bericht*, 41–2.
[65] Maestlin, *Ausführlicher und Grundtlicher Bericht*, 42–4.

him – is interested in the question of the importance of accurately dating festivals and holy days; he too argues that the dates of festivals and holy days are part of ceremonial law and thus not eternally fixed but temporal and subject to change.[66] But Maestlin's view is both more historical and more astronomical. He points out that by the time of the earliest councils, the Julian calendar had already begun to be inaccurate; therefore the dates of major festivals are arbitrary. In any case, it is impossible to know when Christ was really born,[67] or when precisely he died, as can be seen from the different dating of Easter in the Eastern church (indeed, the setting of the date of Easter was in reality a political act[68]). And even when precise information is available, as it is for the deaths of some martyrs, if the festival is to be accurately celebrated the pastor would have to be very precisely trained in astronomy for this to be possible.[69] In fact, the precise day is not important.[70] Corrections to the calendar are neither necessary nor wanted: they would lead only to confusion.[71]

Maestlin then turns to the considerations of the *mathematici, astronomici* and *chronologici*. The last of these, who are concerned with the writing of history, are not interested in a reform of the calendar, because the error which has arisen is *proportionaliter*, and thus does not affect the ordering of historical events.[72] Any change would do nothing but introduce complications into this historical measurement of time. For an astronomer or a mathematician, on the other hand, it really does not matter whether the equinox is in March or in May; they are concerned with accurate observations and all that is necessary is that one should agree how the days are to be counted. Moreover, an attempt to be truly accurate would involve a constant recalculation of the calendar, since no calendar is ever going to be so accurate as to need no correction.[73]

Maestlin's criteria for the adoption of the calendar would thus appear to be based upon a consideration of both its accuracy and its usefulness, and he protests that in the papal bull in which the Gregorian calendar is proposed, these questions of 'common usefulness' are not considered.[74] Like Heerbrand, Maestlin applies Luther's argument that celebrating particular feasts on particular days is not an essential part of the practice of faith, but he does so by appealing to his particular knowledge of mathemat-

---

[66] Maestlin, *Ausführlicher und Grundtlicher Bericht*, 46–7.
[67] Maestlin, *Ausführlicher und Grundtlicher Bericht*, 59.
[68] Maestlin, *Ausführlicher und Grundtlicher Bericht*, 74.
[69] Maestlin, *Ausführlicher und Grundtlicher Bericht*, 60.
[70] Maestlin, *Ausführlicher und Grundtlicher Bericht*, 64.
[71] Maestlin, *Ausführlicher und Grundtlicher Bericht*, 93–4.
[72] Maestlin, *Ausführlicher und Grundtlicher Bericht*, 95–6.
[73] Maestlin, *Ausführlicher und Grundtlicher Bericht*, 96–7.
[74] Maestlin, *Ausführlicher und Grundtlicher Bericht*, 110–11.

ics and astronomy in order to demonstrate the impossibility of achieving sufficient accuracy in these matters.

Accuracy versus authority in calendar reform

Both Heerbrand and Maestlin draw on the concerns raised by Luther's discussion of calendar reform and develop them further, Heerbrand with more emphasis on the theological aspects, and Maestlin by seeking an astronomical basis. Maestlin's status as an astronomer was such that Christopher Clavius felt it necessary to comment on this 'noble mathematician but dreadful heretic' in the preface of his response to Joseph Scaliger.[75] Clavius' response to Scaliger considers questions of accuracy similar to those which had been raised by Maestlin: could the calendar actually deliver what it promised? Maestlin argued that it could not; the Jesuit, Christopher Clavius asserted that it could. But, despite Maestlin's concern to demonstrate that the accuracy of the new calendar could not be sufficient, an assessment of his response which suggests that his arguments are concerned solely with accuracy misses the points of his objections as it does those of Heerbrand. For both Maestlin and Heerbrand, the primary issue is not accuracy, but authority.

Both Maestlin and (albeit to a lesser extent) Heerbrand were convinced that the accurate study of the natural world could lead to a knowledge of God.[76] Both make explicit elsewhere their understanding that because the motions of the heavenly bodies are part of creation, they may offer traces of God's divinity and God's will for the world; as such they belong to the essentials of the faith, and Maestlin will appeal to such arguments in discussing observations which affect his understanding of the structure of the universe. But neither brings to the discussion of the calendar any consideration of time *per se*, or of the regular order of the planets themselves. For the calendar is different: it is not a description of the structure of the universe but a means of measuring or of counting the movements of the planets, and as such is human and not an intrinsic part of the world as created by God. Heerbrand's distinction between the adiaphora and the essential parts of the faith is the theological articulation of the recognition that the calendar is not intrinsic to the year, but is a human way of recording the year. This theological distinction allows both Heerbrand and Maestlin to apply their concerns about the ceding of political authority to the Pope to their decision to reject the new calendar.

---

[75] Clavius, *Josephi Scaligeri Enlenchus*, 4–5.
[76] Maestlin uses this argument to justify his criticism of Aristotelian cometary theory: see Methuen, *Kepler's Tübingen*, 153–7, 171–7.

Part III

Confession and Authority

7

From sola scriptura *to* astronomia nova: *Authority, Accommodation and the Reform of Astronomy in the Work of Johannes Kepler*

Thirty years ago, Jürgen Hübner argued for the placing of Johannes Kepler's theology 'zwischen Orthodoxie und Naturwissenschaft', between natural science (or knowledge of nature) and orthodoxy. Hübner noted Kepler's sense of standing between the confessions, and argued that he drew on a range of theological traditions, most especially Calvinist and Lutheran, in expounding his theological understanding of the world, whilst at the same time maintaining that the Church – despite its divisions – was fundamentally one.[1] Hübner's analysis, whilst profoundly illuminating and of continuing importance, does, however, tend to mask another interpretative concern: to what extent is Kepler's theology best understood as confessional at all? In recent years, church historians have developed an increasingly nuanced understanding of the process of confessionalization, that is, of defining confessional identity, particularly in the later sixteenth and early seventeenth centuries.[2] Much of Kepler's life was led in a context

---

[1] Hübner, *Die Theologie Johannes Keplers*.
[2] For a useful summary of the discussion of the process of confessionalization see Schmidt, *Konfessionalisierung im 16. Jahrhundert*. For specific discussions of the process in Lutheranism, see Rublack (ed.), *Die lutherische Konfessionalisierung*, and for recent insights into the complexity of

in which confessions were not yet fixed but were in the process of becoming more closely and carefully (and, indeed, at times brutally) defined, in terms not only of religious but also of political identity.

In theological terms, Kepler placed himself between confessions, describing himself as a 'catholic' in the truest sense of the word.[3] In this he was standing against contemporary trends. Nonetheless, he was personally affected by a number of aspects of confessionalization.[4] He faced censure on account of his eucharistic theology, which in its understanding of the real presence was Calvinist, or even Zwinglian; in consequence he was excluded from the Eucharist of his own church.[5] Like Philip Apian before him, Kepler found himself unable in conscience to sign the *Formula concordiae*, which was intended to ensure doctrinal unity amongst Lutherans and to which all teachers and pastors in Württemberg were expected to assent.[6] His career was shaped by confessionalization, not least in terms of the several forced moves he made in consequence of the series of expulsions introduced in the course of the Counter-Reformation or re-Catholicization of Styria and the eastern reaches of the empire. And Kepler lived the final decade of his life under the shadow of the Thirty Years' War, which was at root a confessional conflict.[7]

On one level there can be no doubt that there are specific aspects of Kepler's theology, and particularly his views on the Eucharist, which allow him to be placed on the confessional spectrum. However, this chapter will argue that Kepler's explicit placing of himself *between* the confessions reflects his own theological perception of his task: he is engaged on the search for a truth which will reach *beyond* confessional difference and promote harmony where there is discord. Indeed, his understanding of his task

---

the process see the essays in Greyerz *et al.* (eds), *Interkonfessionalität – Transkonfessionalität – binnenkonfessionelle Pluralität*. For the events which led to the re-Catholicization of Styria and Austria, see MacCulloch, *Reformation*, 442–63, or in greater detail, Pörtner, *Counter-Reformation*.

[3] Hübner, *Die Theologie Johannes Keplers*, 101–8. Hübner places Kepler in the context of those who seek to reunite the confessions, although he notes also Kepler's 'antikonfessionalistische Haltung' (107).

[4] For Kepler's biography, see Caspar, *Kepler*; for the theological and ecclesiastical aspects of his life compare also Hübner, *Die Theologie Johannes Keplers*, 2–100. Methuen, *Kepler's Tübingen*, offers a study of the confessional context in which Kepler was educated.

[5] See Hübner, *Die Theologie Johannes Keplers*, 45–60.

[6] For Kepler's refusal to sign the Formula of Concord, see Hübner, *Die Theologie Johannes Keplers*, 108–11. For a detailed discussion of the making of the Formula of Concord and its reception, see Dingel, *Concordia controversa*. Philip Apian, predecessor (and teacher) of Kepler's teacher Michael Maestlin as Professor of Mathematics at Tübingen, had refused to sign and had been dismissed from his post in consequence: see Günther, *Peter und Philipp Apian*, 106–109; and compare Hofmann, *Die Artistenfakultät*, 198.

[7] For the confessional aspects of the development of the Thirty Years' War, see e.g. MacCulloch, *Reformation*, esp. 485–500.

as an interpreter of the heavens – and thus of the Book of Nature as a source of knowledge of divine revelation – is congruent rather with the hopes and ideals of the early Reformers, and particularly Philip Melanchthon, than with the interests of the theologians of his own generation. Kepler termed himself a *reformer of astronomy*; in its methods and aims, the astronomico-theological enterprise in which he was engaged had closer parallels to the methods and priorities of theological reform than to the confessional concerns of his own age. Like Luther and Calvin, Kepler sought to draw on divine authority to establish a truth which he believed could be recognized by all. However, unlike Luther and Calvin, Kepler appears to have asserted the primacy of nature over Scripture in determining the truth about God's creation. In doing so he made use of the theology of accommodation in a way which is closely related to his definition of hypotheses.

Kepler's theological project

Much of the theological energy of the later sixteenth century was dedicated to determining and defining theological distinctions in order to place any given theology confessionally. The theological debates of late sixteenth-century Lutheranism centred on establishing the correct understanding and interpretation of the works of Luther, that is, on the definition of a confessional *regula fidei* which could direct biblical interpretation and with it theology.[8] Although, as Hübner has shown, Kepler did engage with those questions when he had to – particularly in the context of his eucharistic theology – in his justifications of his profoundly theological astronomical work he seems to be attempting something more fundamental.

Kepler consistently pointed to the theological reasons for the study of astronomy. At the beginning of his publishing life, in the preface to the original edition of the *Mysterium Cosmographicum* in 1596, the terms he uses are shaped by the sense that he is studying the heavens to the glory of God:

> Here we are concerned with the book of nature, so greatly celebrated in sacred writings. It is in this that Paul proposes to the Gentiles that they should contemplate God like the Sun in water or in a mirror. Why then as Christians should we take any less delight in its contemplation, since it is for us with true worship to honour God, to venerate him, to

---

[8] For these discussions, see the literature cited in nn. 2 and 4 above.

wonder at him? The more rightly we understand the nature and scope of what our God has founded, the more devoted [devout] the spirit in which that is done.[9]

The study of the heavens reveals God as creator and delights the mind, also created by God.[10] To study the heavens, then, is to know God as creator.[11]

The prefaces to Kepler's *Harmonices Mundi* and the second edition of the *Mysterium Cosmographicum* are marked by his distress at the outbreak of the Thirty Years' War, and by his wish to affirm and demonstrate a vision of God's intended harmony for the world. The *Harmonices Mundi* is dedicated to James VI and I, who, Kepler asserted, had 'removed in the happiest way the hereditary discord between two extremely hostile nations'.[12] Such a removal of discord, which Kepler saw as a demonstration of harmony ('what else is a kingdom but a harmony?' he asked[13]), was, Kepler believed, representative of God's desire for harmonious order, not only in the natural world, but also in society. In terms reminiscent of Philip Melanchthon, who had maintained that the search for the *vestigia Dei* in the natural world must lead the observer to God, Kepler argued that because mathematics and astronomy help observers to recognize the underlying order of the natural world, they also serve as a basis of virtue and orderly moral behaviour and thus of peace. Thus, in the preface to the second edition of the *Mysterium Cosmographicum* (written in 1621, three years after the outbreak of the Thirty Years' War), Kepler recalled Plato's comment that

> according to Apollo's opinion the Greeks turned to geometry and other philosophical studies, as these studies would lead their spirits from ambition and other forms of greed, out of which wars and other evils arise, to the love of peace and to moderation in all things.[14]

In *Harmonices Mundi*, Kepler cited Proclus' commentary on Euclid to remind his readers of the deeper purpose of mathematical endeavour:

> Mental endeavour is the preparation for theology. For those features which to the uninitiated in the truth of divine matters seem difficult

---

[9] Kepler, *Mysterium Cosmographicum*, 53; KGW, vol. 1, p. 5, ll. 24–9; compare KGW 8, 16, 24–9. References to Kepler's works will give first the page number of the English translation, which I have reproduced unless otherwise noted, followed by the volume, page and line numbers in KGW.

[10] Kepler, *Mysterium Cosmographicum*, 53–5; KGW 1, 5–6; compare KGW 8, 16–17.

[11] Very similar justifications for astronomical work, with references to the same biblical passages, are found also in prefaces to works by Michael Maestlin, and it is likely that Kepler learned this approach from him. See Methuen, *Kepler's Tübingen*, 152–8, 173–82.

[12] Kepler, *Harmonices Mundi*, 4; KGW 6, 10, 27.

[13] Kepler, *Harmonices Mundi*, 4; KGW 6, 10, 25–6.

[14] Kepler, *Mysterium Cosmographicum*, 43; KGW 8, 11, 31–5.

to grasp and lofty are by mathematical reasoning shown to be trustworthy, manifest and uncontroversial, by means of certain images. . . . Thus Plato teaches us many remarkable things about the nature of the gods through the appearance of mathematical things; and the Pythagorean philosophy disguises its teaching on divine matters with these, so to speak, veils. . . . Again, it perfects us in moral philosophy, implanting in our behaviour order, propriety and harmony in social relations. . . . By all of these things we are guided to the middle way in behaviour and in morals.[15]

As an astronomer and a mathematician – a priest of the Book of Nature, as he himself described his task – Kepler saw himself as having access to an understanding of divine truth which was ultimately capable of reforming human society, transcending confessional differences and leading to peace. For him, this was the ultimate reason for engaging in the study of astronomy, and his theological astronomy was therefore intended to transcend confessional difference and indeed to offer a way of uniting the peoples of the world, just as Scotland and England had been united under James VI and I. Kepler, then, was not seeking to define himself theologically in terms of confessional theology; rather, like the first generation of Reformers, he was explicitly concerned with a direct discovery of God's truth and its implications for the world. For him, mathematics and astronomy offered one way of accessing revealed truth, and, as will be shown below, in some areas of understanding, Kepler saw them offering a more certain path to truth than that offered by the interpretation of Scripture.

Ancient authorities and the reform of astronomy

Although for Kepler astronomy could offer a means to deeper perception of theological truth, it was unable to do so in its current state. Astronomy, like religion and the Church before it, must therefore be reformed. The poem to the reader which introduces the 1596 edition of *Mysterium Cosmographicum* speaks of Kepler's discovery of the secret contained within it in terms of rebirth: 'Clearly [God] has revealed by this example that we can be reborn (*nos posse renasci*) after two thousand years.'[16] The language of rebirth clearly alludes to salvation,[17] and the reference to rebirth after two thousand years also carries overtones of the message of the Reformers.

---

[15] Kepler, *Harmonices Mundi*, 127–8; KGW 6, 95.
[16] Kepler, *Mysterium Cosmographicum*, 49; KGW 1, 4; cf. KGW 8, 14.
[17] For instance, for Calvin the saved are the 'reborn', the *renati*.

In the introduction to the 1621 edition of the *Mysterium Cosmographicum*, in a note of thanks to those who had supported and continued to support his work, Kepler reflected on what he had been doing since the first edition a quarter of a century earlier:

> Throughout these last twenty-five years, while I have been weaving the fabric of the reform of astronomy (started by Tycho Brahe of the Danish nobility, the very celebrated astronomer), they have carried a torch before me more than once.[18]

His project amounts to the 'reform of astronomy', *restauratio astronomiae*, which means more properly the *restoration* or *renewal* of astronomy. Kepler presents his work in terms entirely commensurate with the humanist programme: the *reform* or *renewal* of astronomy is a restoration, in the sense of a recovery of ancient, pure truth; indeed, he implies, of a truth which has been lost for two thousand years. Kepler is working within a context in which novelty is suspect, or even heretical.[19] His rhetoric of referring back to a truth recognized in ancient times is similar to that of the Reformers, who had described their reforms in terms of a return to the true and ancient church and the stripping away of intervening accretions of theology and practice. Kepler's description of his work as *restauratio astronomiae* suggests that the expectation that novelty will pervert rather than reveal truth is still a factor in his thinking.[20]

That this concern can be traced also in the preface to the *Astronomia Nova* may seem somewhat paradoxical, and it is indeed in this work that a subtle shift in his appeal to antiquity becomes apparent.[21] The care with which Kepler sought to establish the antiquity of his ideas is apparent in his consideration of the different schools of astronomy.[22] He identifies two schools of thought: that of 'Ptolemy and a large majority of the ancients', which he regards as inferior for its treatment of the planets independently from one another, and another 'attributed to more recent proponents, although it is the most ancient', which 'relates planets to each other, deduces from a common cause those characteristics found to be common

---

[18] Kepler, *Mysterium Cosmographicum*, 39; KGW 8, 9, 22–5.

[19] The suspicion of novelty, and the conviction that what was new was heretical, led the founders of new traditions to search for precedents in antiquity. For a discussion of these questions, centred on the term *inventio* and its shifts in meaning, see Atkinson, *Inventing Inventors*, 14–66.

[20] See William H. Donahue's remarks in Kepler, *Astronomia Nova (The New Astronomy)*, Translator's Introduction, 1–2.

[21] Kepler tended to refer to this work by its subtitle, *Commentaries on Mars*, perhaps in part an expression of his unease with the implication of novelty.

[22] Kepler, *Astronomia Nova*, 47–9; KGW 3, 19–20. Kepler's phrase *De sectis astronomorum* is translated by Donahue's 'On the schools of thought in astronomy'. Although a legitimate translation, this misses the allusion to sects in the ecclesiastical sense.

to their motions'.²³ This latter school, Kepler explains, is now subdivided: 'Copernicus with Aristarchus of remotest antiquity ascribes to the translational motion of our home the earth the cause of the planets' appearing stationary and retrograde', whilst Tycho Brahe ascribes this cause to the sun.²⁴ The implication is that Copernicus's hypothesis, supported by Aristarchus as the most ancient authority, is the more reliable.

However, Kepler is also careful to point out that his judgement is based not only on ancient authorities, but on an evaluation of the observational evidence. His aim, he writes, has been

> to amend astronomical theory (especially of the motion of Mars) in all three forms of hypotheses, so that our computations from the tables correspond to the celestial phenomena.... I also made an excursion into Aristotle's *Metaphysics,* or rather I enquired into celestial physics and the natural causes of the motions. The eventual result of this consideration is the formulation of very clear arguments showing that only Copernicus's opinion concerning the world (with a few small changes) is true, that the other two are false, and so on. Indeed, all things are so interconnected, involved, and intertwined with one another that after trying many different approaches to the amendment of astronomical calculations, *some well trodden by the ancients and others constructed in emulation of them and by their example,* none other could succeed than the one founded on the motions' physical causes themselves, which I establish in this work.²⁵

Kepler's use of and appeal to ancient authority seems here to be changing to allow him to define consistency with the ancients in terms not of their knowledge, or even of their specific theories and hypotheses, but in terms of their *method* of approaching the problem, and the measurable results produced by that method. Kepler does not simply accept the teachings of even those ancient authorities with whose interpretations he concurs. Instead he goes a step further, regarding 'emulation' – structuring his work according to their example – as a legitimate means of following ancient authorities, and thus of protecting himself against charges of novelty.

Kepler is clearly treading a careful line here between ancient authority and accusations of innovation. He is particularly sensitive to the possibility of accusations of innovation in cases where older methods – methods that

---

[23] Kepler, *Astronomia Nova,* 47; KGW 3, 19, 25–9.
[24] Kepler, *Astronomia Nova,* 47; KGW 3, 19, 29–40.
[25] Kepler, *Astronomia nova,* 48; KGW 3, 20, 4–7 (my italics, translating *partim a veteribus tritis, partim ad eorum imitationem et exemplum structus*). 'Amend' at the beginning of this quotation and 'amendment' in line 11 are my translation. Donahue translates the Latin *emendare* as 'reform', which, particularly in the context of this discussion, is potentially misleading.

might be seen as traditional – are able to explain his results. Thus, in his discussion of the first step of his argument, the question of the point of intersection of the eccentrics of the planet's motion, which he places in the centre of the sun, Kepler notes that 'here the Braheans could have raised the objection against me that I am a rash innovator', since their calculations also correspond to what is observed.[26] The Ptolemaic system also accounts for the phenomena. 'Therefore,' he comments, 'I have to look again and again at what I am doing, so as to avoid setting up a new method which would not do what was already done by the old method.'[27] From this point in Kepler's argument, his own methods are opposed to 'the old methods' (and thus tacitly accepted to be new), and demonstrated to give simpler explanations which better accord with the observations. For Kepler, the advantage of his methods is that they allow one physical theory to account for the motion of the universe using only one kind of motion: the hypothesis of Copernicus has 'entirely removed [the] extrinsic motion from the planets, assigning its cause to a deception arising from the circumstances of observation. Thus the motions are still multiplied to no purpose in Brahe, as they were before by Ptolemy.'[28] Simplicity is a decisive factor in allowing the introduction of what some take to be innovation.

It is clear that Kepler is here working hard to define and assess a variety of ways of judging between the different astronomical 'hypotheses' with which he is confronted. On one level, he is developing 'a methodology in which hypotheses are built on and confirmed by observations',[29] but in doing so he is concerned both to establish that his methodology has ancient antecedents and to establish a means by which both ancient authorities and their modern interpreters can be judged when they disagree. Additionally, he is seeking to understand more deeply God's plan of creation; it is his quest for what he understands to be the divinely imparted structure which leads him to insist that an explanation must be both mathematical and physical, accounting at one and the same time for appearances and causes.

The structure of Kepler's 'reform of astronomy' has similarities to the rhetorical structure of the Reformation, particularly as it was perceived by the Reformation's opponents. Thus Kepler insists on the congruence of his astronomical reforms with ancient authorities, and with Plato and Pythagoras in particular. At the same time, he is conscious of offering methods for distinguishing between different hypotheses, which he recognizes are not the same as the 'old ways'. Structurally, this procedure reflects

---

[26] Kepler, *Astronomia nova*, 49; KGW 3, 20, 32–3.
[27] Kepler, *Astronomia nova*, 49; KGW 3, 20, 39–41.
[28] Kepler, *Astronomia nova*, 51; KGW 3, 23, 10–11.
[29] Kepler, *The Harmony of the World*, Translators' Preface, viii.

many of the criticisms levelled at the methodology of the Reformers, who were accused of claiming ancient authority for theology and practices which were in fact innovatory and therefore heretical.[30] For the Reformers the measure of those practices was the Bible, which they considered to reveal the mind of God and thus theological truth. Kepler's measure of the truth of his astronomical theories, however, is not Scripture. Rather, he conceives his work of the restoration of astronomy to be measured against knowledge of the universe created by God. The astronomy that reveals the truth about the created order of the universe must be the true astronomy, and by definition can be no novelty.

Scripture and the knowledge of the universe

Kepler maintains that the truth of his methods must be judged according to the authority of reason, based on observation of the created world, and not simply by reference to the knowledge or observations of ancient authorities. Amongst these authorities is Scripture, for whilst Kepler believes that the Bible must be understood to be the measure of piety and of theological truth, he is adamant that it cannot explain the physical truths of the universe, for it does not teach knowledge of the universe. 'Read all of Chapter 38 in Job, and compare it with matters discussed in astronomy and in physics,'[31] he exhorts. What pervades Scripture is the conviction that God created the world: 'This whole worldly edifice that you see, light above and dark and widely spread out below, upon which you are standing and by which you are roofed over, has been created by God.'[32] That act of creation calls humankind to worship, and it is that worship – and not particular arguments from natural philosophy – which is the concern of scriptural texts concerning the natural world.

In considering the purpose of Scripture, Kepler therefore makes a clear distinction between the methodologies of theology and those of astronomy:

> As for the opinion of the pious on these matters of nature, I have just one thing to say: while in theology it is authority that carries the most weight, in philosophy it is reason. Therefore, Lactantius is pious, who denied that the world is round, Augustine is pious, who, though admitting the roundness, denied the antipodes, and the Inquisition

---

[30] Note, for instance, how Calvin defends himself against charges of novelty in theology and ordering of the Church: Calvin, *Responsi ad Sadoleti epistolam*, CO 5, 392–6; *Reply to Sadolet*, 229–34.

[31] Kepler, *Astronomia nova*, 62; KGW 3, 30, 40–1.

[32] Kepler, *Astronomia nova*, 62; KGW 3, 20, 29–31.

nowadays is pious, which, though allowing the earth's smallness, denies its motion.[33]

Despite his affirmation of their piety, Kepler is convinced that Augustine and Lactantius would be more pious still if they were to recognize the truths of the natural world: 'to me [writes Kepler] the truth is more pious still'.[34] However, piety is not dependent on natural knowledge, and ignorance in matters of the natural world is not a bar to the worship of God, particularly in the case of those who have neither the opportunity nor the understanding for knowledge of the created world. At the same time, those who are in this position should simply praise God and not seek to instruct those who have deeper (or, in Kepler's view, higher) knowledge:

> To whoever is too stupid to understand astronomical science, or too weak to believe Copernicus without affecting his faith, I would advise him that, having dismissed astronomical studies and having damned whatever philosophical opinions he pleases, he mind his own business and betake himself home to scratch in his own dirt patch, abandoning this wandering about the world. He should raise his eyes (his only means of vision) to this visible heaven and with his whole heart burst forth in giving thanks and praising God the Creator. He can be sure that he worships God no less than the astronomer, to whom God has granted the more penetrating vision of the mind's eye, and an ability and desire to celebrate his God above those things he has discovered.[35]

Rather than condemning the ignorant, God allows them to worship in accordance with their understanding and their faith; nonetheless the astronomers' work reveals higher truths.

Kepler's 'advice for the common people' – as he himself terms this passage[36] – cannot conceal his conviction that a true understanding of Scripture will lead the believer to recognize Scripture's limitations in questions of natural philosophy. He demonstrates this by showing that Psalm 104 is rightly to be understood as a commentary on the creation narrative in Genesis 1 and not as a treatise in natural philosophy.[37] Moreover, Kepler notes that many passages of Scripture use figures of speech or simply recount what is seen from the observer's point of view. As an astronomer, he knows and respects the fact that perspective can make a simple motion look different and emphasizes that this is a perfectly normal aspect of living in the world:

---

[33] Kepler, *Astronomia nova*, 66; KGW 3, 33, 37–44, 2.

[34] Kepler, *Astronomia nova*, 66; KGW 3, 34, 2–5.

[35] Kepler, *Astronomia nova*, 65–66; KGW 3, 33, 17–26.

[36] Kepler, *Astronomia nova*, 65; KGW 3, 33, 27 (*consilium pro idiotis*).

[37] Kepler, *Astronomia nova*, 63–5; KGW 3, 31–3. For a detailed discussion of Kepler's exegesis of Psalm 104, see Howell, *God's Two Books*, 121–5.

to absolutely all men, the sun appears to move and not the earth.... It is therefore impossible for a previously uninformed reason to imagine anything but that the earth, along with the arch of heaven set over it, is like a great house, immobile, in which the sun, so small in stature, travels from one side to the other like a bird flying in the air.[38]

Such questions are not relevant to the central message of Scripture, which is theological and moral, concerning the mutability of humankind and the constancy of God's created universe.[39] In order to purvey the message of salvation, Scripture needs to be able to speak to the ignorant as well as to the learned, and so the images which it uses must be comprehensible to the simplest mind. This, however, must not deter the astronomer from seeking to know the truth about God as revealed in the heavens.

Kepler's understanding of the relationship between the Books of Scripture and Nature seems to have developed alongside his knowledge of astronomy. However, this development is difficult to trace, for although it is known that a discussion of the authority of Scripture was omitted from the first edition of the *Mysterium Cosmographicum* at the recommendation of Michael Maestlin, it is impossible to say whether those omissions were identical with the discussion subsequently published in the *Astronomia Nova*. As noted above, in the 1596 edition of the *Mysterium Cosmographicum*, Kepler argues, as Maestlin had before him, that true contemplation of the heavens – which is to say, accurate study of them – is done to the glory of God. He recognizes, however, that 'these heavenly matters are not nourishment for everyone indiscriminately, but only for a noble mind'.[40] The study of the heavens is not human but divine, and will appeal to the man who

> has fully explored all possibilities, yet, as these things are human, he has found nothing anywhere which is blessed with happiness, everlasting, and able to satisfy and satiate his appetites. For then he will begin to seek better things, then he will ascend from Earth below to heaven, ... therefore he will begin to despise what once he thought most important, he will value only these works of God, and he will derive pure and sincere delight at last from these studies.[41]

---

[38] Kepler, *Astronomia nova*, 62; KGW 3, 30, 17–25.
[39] Kepler, *Astronomia nova*, 63; KGW 3, 31–2.
[40] Kepler, *Mysterium Cosmographicum*, 55; KGW 1, 7, 7–10. Compare KGW 8, 18, 10–13.
[41] Kepler, *Mysterium Cosmographicum*, 55–7; KGW 1, 7, 12–23. Compare KGW 8, 18, 15–27.

The study of astronomy therefore offers a counter to the 'profit, wealth, and treasures' of the world.[42] Kepler affirms (in terms reminiscent of Luther's stance at the Diet of Worms 75 years before): 'I promise generally that I shall say nothing that would be an affront to Holy Scripture, and that if Copernicus is convicted of anything along with me, I shall dismiss him as worthless.'[43]

Reflecting on his own words in 1621, Kepler expanded on the distinction between the Book of Scripture and the Book of Nature which he had discussed in the *Astronomia Nova*, expressing his hope that he had been able there

> to have satisfied those with religious scruples, provided that they approach their decision on this point with sufficient intelligence and knowledge of astronomy for the glory of God's works, which are themselves visible, to be safely entrusted to our protection.[44]

For Kepler, the distinction between the Books of Scripture and Nature reveals that they are intended by God to witness to different aspects of God's nature and work.

The distinction between the two books thus reveals important aspects of the relationship between knowledge drawn from theological insights and that drawn from astronomy or natural philosophy. Kepler argues at some length that on matters of natural philosophy and astronomy, the revelation found in the Book of Scripture cannot be taken to contradict the Book of Nature:

> Certainly God has a tongue, but he also has a finger. And who would deny that the tongue of God is adjusted both to his intention, and on that account to the common tongue of men? Therefore in matters which are quite plain everyone with strong religious scruples will take the greatest care not to twist the tongue of God so that it refutes the finger of God in Nature. Let him read, if any man is concerned for the praises of our Creator and Lord, let him read, I say, the fifth book of my *Harmonices*, and when he has perceived the most skilfully harmonized Republic of the motions, let him debate with himself whether sufficiently sound, sufficiently prolific reasons have been discovered for reconciling the tongue and finger of God; or whether he will repudiate that reconciliation and hasten to suppress with

---

[42] Kepler, *Mysterium Cosmographicum*, 57; KGW 1, 7, 24–5. Compare KGW 8, 18, 28–9. There are overtones here of Luther's critique of justification by works and of a theology of glory, although as argued in Chapter 2, Luther was extremely cautious about the ability of human reason to know God.

[43] Kepler, *Mysterium Cosmographicum*, 75; KGW 8, 31, 7–9.

[44] Kepler, *Mysterium Cosmographicum*, 85; KGW 8, 39, 30–2.

censorship the renown of the immeasurable splendour of the works of God. That this renown should come to be known to the common people, nay rather, to the generality of the even superficially educated, could never be brought about by order. Ignorance refuses to respect authority; it resorts spontaneously to combat, relying on numbers and on the shield of habit, which is impenetrable to the weapons of truth.[45]

Here Kepler seems to offer a rather less sanguine assessment of the situation for those who are ignorant of astronomy than his earlier 'advice to the common people'. More importantly, his discussion reveals a striking principle: once an astronomer has recognized the truth about it, the Book of Nature does not have to be – and indeed cannot be – accommodated or adjusted to the understanding of the ignorant, whereas the Book of Scripture can and must be. Kepler is here drawing on the theory or theology of accommodation, but there is a twist to his use of it. Precisely because the 'finger of God', which is nature, does not need to be adapted to the incapabilities of the ignorant, it can reveal some aspects of the creator God more clearly; the 'tongue of God', on the other hand, which is Scripture, has a message of salvation which must be capable of being communicated to and understood by everyone. It therefore has to use a language – even about the natural world – which makes such understanding possible.

The idea that Scripture is accommodated to human capacities is an ancient one, which had been taken up by the Reformers, and in particular by John Calvin.[46] Whilst Calvin believed that the created world displays God's wonderful wisdom on different levels, 'not only those more recondite matters for the closer observation of which astronomy, medicine and all natural science are intended, but also those which thrust themselves upon the sight of even the most untutored and ignorant', and that God's providence 'shows itself more explicitly' to those who have studied the liberal arts or astronomy,[47] he also emphasized that although 'the invisible divinity is made manifest' by nature (according to Romans 1.19), 'we have not the eyes to see this unless they be illumined by the inner revelations of God through faith'.[48] Fallen human nature and reason tends to use its knowledge to set up its own theories rather than glorifying God: 'at the same time as we have enjoyed a slight taste of the divine from contemplation of the universe, having neglected the true God, we raise up in his stead dreams and spectres of our own brain'. In Scripture 'God is wont in a measure to "lisp"

---

[45] Kepler, *Mysterium Cosmographicum*, 85; KGW 8, 39, 32–40, 4.
[46] For a discussion of Calvin's view of accommodation, see Forstman, *Word and Spirit*, and compare Benin, *Footprints of God*, 187–95.
[47] Calvin, *Institutio*, 1, 5, 2 (LCC 20, 53); English translations in the text are from LCC.
[48] See, for instance, Calvin, *Institutio*, 1, 5, 14 (LCC 20, 68).

in speaking to us', so that 'such [scriptural] forms of speaking do not so much express clearly what God is like as accommodate the knowledge of him to our slight capacity'.[49] Therefore, he argues in his commentary to Genesis 1.16, 'Moses is not analyzing accurately, like the philosophers, the secrets of nature'.[50] Indeed, 'when the astronomer seeks the true size of stars and finds the moon smaller than Saturn, he gives us specialized knowledge. But the eye sees things differently and Moses adapts himself to the ordinary view'.[51] Scripture is therefore accommodated to ordinary human capabilities. This is not to decry the use of astronomy:

> Moses described in popular style what all ordinary men without training and education perceive with their ordinary senses. Astronomers, on the other hand, investigate with great labor whatever the keenness of man's intellect is able to discover. Such study is certainly not to be disapproved, nor science condemned with the insolence of some fanatics who habitually reject whatever is unknown to them. The study of astronomy not only gives pleasure but is also extremely useful. And no one can deny that it admirably reveals the wisdom of God.[52]

However, that 'revelation of the wisdom of God' can be no more than a glimpse of the divine nature. Although astronomers can prove their assertions 'by conclusive reasons' and their knowledge points to God, Calvin's understanding of the Fall and the consequent dimming of human reason means that a true grasp of the nature of God through the natural world is not possible to human reason; reason must be complemented by Scripture.[53] For Calvin, then, both Scripture and nature reveal God in a way that is accommodated to fallen human nature, whether educated or not. To comprehend the divine revelation in nature requires more informed knowledge and greater use of reason than to comprehend the divine revelation in Scripture; nonetheless nature is not intrinsically truer to God than is Scripture.

Unlike Calvin, Kepler appears convinced that through mathematics and astronomy the natural world reveals to the learned the true structures of the universe, which in their turn reveal the true nature of God as Creator. This is a deeper, purer knowledge than that revealed by Scripture to the ignorant and untutored, that is, to those who have not (yet) grasped the principles of relative motion.

---

[49] Calvin, *Institutio*, 1, 13, 1 (LCC 20, 121).
[50] Calvin, *Commentarius in Genesin*, 22; CR 51/CO 23, 22.
[51] Calvin, *Commentarius in Genesin*, 22; CR 51/CO 23, 22.
[52] Calvin, *Commentarius in Genesin*, 22–3; CR 51/CO 23, 22.
[53] See the extended discussion in Calvin, *Institutio*, 1, 5 (LCC 20, 62–9). On the need for scriptural revelation to complement knowledge of the natural world, see Calvin, *Institutio*, 1, 6, 1 and 4 (LCC 20, 69–70, 73–4).

Hypotheses, appearances and accommodation

The questions that Kepler raises in his discussion of accommodation are strikingly close to the questions which concern him in his discussion of hypotheses. As noted above, one criterion adopted by Kepler as a measure of the congruence of mathematical and physical explanations is simplicity, which he sees in terms of whether the mathematical hypothesis correlates with the physical reality of a particular motion rather than simply explaining its appearance from the point of view of an observer on the earth. In applying this criterion, he makes a central distinction between 'fictitious hypotheses' and 'astronomical or true hypotheses'. The former are those which propose a mathematical explanation for the observed motion and nothing more; the latter explain the motion but at the same time offer an account of the physical reality:

> If an astronomer says that the path of the moon is an oval, this is an astronomical hypothesis representing the true motion. When, however, he proposes a combination of circular motions by which the oval orbit may be described, he is proposing a geometric or fictitious hypothesis.[54]

A 'geometric or fictitious hypothesis', such as that of Ptolemy or Brahe, can offer a mathematical explanation for what is observed, but bears no physical relation to what is actually happening in the heavens. The Copernican hypothesis, on the other hand, explains both, and therefore reveals more about the structure of the heavens. This distinction is very similar to that which Kepler makes between the tongue of God, which in matters of natural philosophy and astronomy speaks through Scripture and must accommodate itself to the knowledge and situation of its readers, and the finger of God, which reveals the truth of the structure of the universe – a truth which Kepler believes to be identical with that revealed by his 'true hypothesis'. Kepler's understanding of the theological principle of accommodation and its relationship to the nature of God can thus be seen as closely related to his philosophical-methodological deliberations on the question of hypotheses.

---

[54] Kepler, *The Harmony of the World*, Translators' Introduction, xii–xiii; quote from p. xii.

Kepler: a Lutheran astronomer?

It is clear from the above that Kepler's understanding of the principle of accommodation shows a marked shift from Calvin's use. Kepler is clear that Scripture is not the primary measure of truth when it comes to judging the accuracy of physical or astronomical – or physico-astronomical – theories. It is the 'finger of God' rather than the 'tongue of God' that reveals God's true nature and as such is to be believed. Kepler is beginning to distinguish between scriptural and theological truth, as well as between scriptural and philosophical or physical truth, a step which had already been taken by his teacher Michael Maestlin. It has been observed that the Reformation saw an expansion in the remit of Scripture,[55] but this expansion is curtailed by Kepler's approach to accommodation.

Howell has shown how Kepler's use of accommodation emphasizes the scriptural texts as historical and mutually dependent. This development is underlined by the parallels between Kepler's use of accommodation and his approach to hypotheses, which implicitly place Scripture on the level of an unsubstantiated, untutored hypothesis when it comes to matters of astronomical observation. Kepler sees the texts of Scripture as in some way historically conditioned.

At the same time, Kepler's understanding of the task on which he is engaged – that of the reform of astronomy – has much in common with the way that the Reformers conceived their task. Kepler's aim is to uncover the truth of the universe and to free that truth from the accretions of those who have held themselves to be astronomers, just as, almost a century before, Luther had sought to free the Church and theology from the accretions of those held to be theological or ecclesiastical authorities. Kepler's strong conviction that the study of nature would yield insights not only in natural philosophy but also in moral philosophy, insights which would offer a key to understanding the mind of God, stands directly in the Stoic tradition as mediated by Philip Melanchthon. By the criterion of that approach – that the Reformation would bring peace and civic harmony – the Reformation must be said to have failed. Kepler's appeal to authority – to the ancient, classical authorities of the humanist *ad fontes* – is not consonant with the programme of confessionalization, which he sees as bringing about only strife, but instead with an earlier, almost Erasmian vision of peace through learning.

'I am a Lutheran astrologer, throwing away the nonsense and keeping the kernel', wrote Kepler of his attempt to free astrology from superstition

---

[55] McGrath, *Intellectual Origins*, 151.

(another task dear to Melanchthon's heart).[56] 'I am a Lutheran astrologer.' The term 'Lutheran' here is not to be taken in its late sixteenth-century, confessional sense. Rather, Kepler is Lutheran because he is a Reformer in Luther's model. But Kepler was a Reformer, not of the Church, but of astronomy and astronomical method. He was a Reformer whose reforms were to be measured by the *liber naturae* rather than the *liber scripturae*. He was a Reformer whose understanding of *sola scriptura* ended with *sola natura* – and who, despite his appeals to ancient authority, was aware that he had produced an *astronomia nova* and was prepared to justify that break with tradition.

[56] Letter from Johannes Kepler to Michael Maestlin, 15 March 1598, KGW 13, 184, 177–8; cited as quoted in Field, 'A Lutheran Astronomer', 220.

8

On the Problem of Defining Lutheran Natural Philosophy

One of the most intriguing questions (at least for a theologian) raised by the encounter between the history of theology and the history of science in recent years is whether there can be said to be a confessional approach to natural science. In many ways, this question can be dated back to Robert Merton's thesis that the ethos of English Puritanism was significant for the emergence of science.[1] Merton's thesis was confined to developments in England, but the last few years have seen increasing numbers of scholars turning their attention to the question of the development of natural sciences in the context of the continental Reformation, and in particular to the question of the extent to which it is possible to speak of a Lutheran natural philosophy, or a Lutheran response to Copernicus. Further, if it is, what does the term 'Lutheran' imply? Is it simply a reflection of the fact that the practitioners under consideration came from Lutheran circles, particularly the circles around and succeeding Melanchthon in Wittenberg, or is there something fundamentally confessional, and specifically Lutheran, in their approach?

The question arises from the recognition that in the history of theology, the use of the term 'Lutheran' presupposes the process of confessionalization (defining and distinguishing between Catholicism and Protestantism, or into the Catholic, Lutheran and Calvinist confessions[2]), that is, it tacitly assumes that there are other approaches which may be labelled 'Catholic' or

---

[1] Merton, *Science, Technology and Society*. For a summary of criticisms of Merton's thesis and a discussion of the influence of English Latitudinarianism, see Henry, 'The Scientific Revolution in England'; compare also the essays collected in Kroll *et al.* (eds), *Philosophy, Science and Religion in England* and in Force and Popkin (eds), *The Books of Nature and Scripture*.

[2] For the move from a two-fold to a three-fold model of confessionalization, see Wallmann, 'Lutherische Konfessionalisierung', esp. 37–9.

'Calvinist'. Because the process of confessionalization took some time, the phase of Lutheran differentiation from Calvinism being focused on the period between the Peace of Augsburg in 1555 and the formulation of the *Formula concordiae* in 1570, some of the specific characteristics of Lutheran theology and practice developed over time, with views which had earlier been accepted later being condemned.[3] The definition of what is Lutheran is thus complex. Moreover, within the historiography of the Reformation, discussions of the process of confessionalization, often rooted in social history, have generally sought to analyse differences in practice. Of course, these are often rooted in theology and here eucharistic theology is vital, for it was controversies over the doctrine of the Eucharist which led to the insoluble differences between the Protestant parties. Until now, however, differences in natural philosophy or in approaches to the natural world have not been regarded as criteria for defining confessional difference.[4] This raises an interpretative question: does the refusal to regard natural philosophy as an instance of confessionalization represent a failure to take account of the results of confessionalization upon both the curricula and content of university teaching and the observation of the natural world? Or do claims for a Lutheran approach to nature represent a failure to take account of inter-confessional differences and intra-confessional similarities in the interaction between theology and philosophy?

Observation of the natural world

A central element of the discussion of the role of confession in defining attitudes to the natural world has been the recognition of the importance of what Robert Westman has called 'the Wittenberg circle' in responding to the Copernican hypothesis.[5] It is likely that this interest in the natural world can be traced back to the influence of Philip Melanchthon: Sachiko Kusukawa has highlighted Melanchthon's interest in contemporary discussions of astronomy and of anatomy in his reflections on natural philosophy and the soul.[6] By referring to empirical studies and not only to the texts of Aristotle, Ptolemy, Galen and others, she argues, Melanchthon introduced into his natural philosophy an interest in the study of the natural world as it

---

[3] For a summary of this research, see Wallmann, 'Lutherische Konfessionalisierung', and compare Schmidt, *Konfessionalisierung im 16. Jahrhundert*. For the role of the universities, including the Jesuit academies, in this process see, for instance, Harris, 'Confession-Building'; Methuen, 'Securing the Reformation'; Seifert, 'Das Höhere Schulwesen'.

[4] See, for instance, Brooke *et al.* (eds), *Science in Theistic Contexts*.

[5] Westman, 'The Melanchthon Circle'.

[6] Kusukawa, *The Transformation of Natural Philosophy*; for Melanchthon's use of anatomy, see also Nutton, 'Wittenberg Anatomy', esp. 16–26.

actually is; Kusukawa suggests that in doing so, he established a new, 'Lutheran' trend in natural philosophy.[7] In appealing to Melanchthon's use of the results of anatomical dissections, Kusukawa's thesis may be seen in parallel to that of Andrew Cunningham, who regards such interest in the observation of the natural world as 'Protestant' to the extent of referring to the sixteenth-century Italian anatomist Andreas Vesalius (who was almost certainly a Catholic) as a 'Protestant anatomist'. In justification, Cunningham argues that both Vesalius and Luther appealed to an authority beyond the works generally accepted as authoritative, in particular by taking 'the text of the Bible or the text of the body [to be their] sole authority'.[8] For Cunningham, then, the dissecting anatomists' concern to understand the physical reality of the human body may be viewed as parallel to the Lutheran appeal to the text of the Bible, and thus as Protestant. For Kusukawa, the use of such knowledge in natural philosophy justifies labelling it Lutheran.

It is certainly plausible to see the quest for a new understanding of Scripture and the attempt to observe the natural world more precisely in parallel. Throughout the history of Christianity, theologians have struggled to clarify the relationship between God's two books, that is, the Book of Scripture and the Book of Nature, and their status as sources of knowledge of divine revelation. Focusing particularly on changes in the approach to natural history during the sixteenth century, Peter Harrison has argued persuasively that

> the historical origins of two of the hallmarks of modernity – the identification of the meaning of a text with its author's intention, and the privileged status of scientific discourse – were closely intertwined. The modern approach to texts, driven by the agenda of the reformers and disseminated through Protestant religious practices, created the conditions which made possible the emergence of modern science.[9]

---

[7] Kusukawa, *The Transformation of Natural Philosophy*.

[8] Cunningham, 'Protestant Anatomy', 96–8 (quotation at 97). See also, however, Helm's arguments against a distinctively Protestant anatomy (Helm, 'Religion and Medicine'). For the complexities of the relationship between the natural world (in this case the heavens) and Biblical interpretation in the work of Lutheran astronomers, see Granada, 'Il Problema Astronomico-cosmologico'; Howell, *God's Two Books*; Methuen, *Kepler's Tübingen*, esp. 151–7. It should, however, be noted that Galileo was as keen as Maestlin or Brahe to maintain the authority of Scripture.

[9] Harrison, *The Bible, Protestantism, and the Rise of Natural Science*, 266. Howell's work demonstrates the range of hermeneutical approaches to which this intertwining gave rise in Europe: see especially his *God's Two Books*. Although Howell considers that '[Harrison's] fundamental thesis fails to account for the diversity of hermeneutical approaches' that can be observed in the sixteenth and early seventeenth centuries (ibid., 8), the work of both Howell and Harrison underlines the deep interest in Biblical interpretation held by many of those who observed the natural world. This interest continued: seventeenth-century English scientists such as Robert Hooke and Isaac Newton were greatly interested in the exegesis of Scripture. See Poole, 'The Genesis Narrative' and Iliffe, 'Friendly Criticism'.

Harrison believes that this entwining resulted, at least in part, from a shift of focus in the Protestant Reformation from a symbolic to a more literal understanding of both words and objects.[10] It also reflected the new knowledge gained from the exploration of the 'New World' and the corresponding recognition that Aristotle's knowledge and depiction of the world was circumscribed and incomplete.[11] This recognition, coupled with a profound distrust of scholastic philosophy and theology, led to a crisis of authority and forced both theologians and philosophers to seek more certain ways of establishing proven facts. While philosophers turned to the evidence of their own eyes, or to new methodologies and methods of proof, to achieve this aim,[12] a number of theologians explicitly developed the 'two books' theology.[13]

As might be expected from the course of the 'scientific revolution' and the associated developments in scientific methodology, Harrison concludes that it was not until the seventeenth century that this relationship found its final shape. Throughout the sixteenth century, he argues, under the influence of humanism, natural history saw a progression from the attempt to arrive at an authoritative, uncorrupted version of an original, usually ancient text, through the use of nature to resolve conflicts between such texts or to correct ancient texts, to the collection of 'natural objects' and the attempt to observe the habits and behaviour of animals.[14] Such observations, he suggests, did not at that time lead to the postulation of new modes of classification – for 'critical humanist scholarship . . . was at best parasitic on prevailing paradigms'[15] – they nevertheless raised questions about the

---

[10] Harrison, *The Bible, Protestantism, and the Rise of Natural Science*, 107–20. This is not to suggest that all scholars moved away from the idea that the universe might be understood symbolically: some, such as Johannes Kepler, clearly continued to believe this. See Howell, *God's Two Books*, 8.

[11] Harrison, *The Bible, Protestantism, and the Rise of Natural Science*, 83–4, and compare Methuen, 'Maestlin's Teaching of Copernicus', 236 n. 24.

[12] Increased certainty was the stated aim of most of these new methodologies: see, for instance, the work of Ramus ('Ramus, Petrus', in TRE 28, 129–33) and Grynaeus (Methuen, *Kepler's Tübingen*, 165–71). Kusukawa implies that this aim was particularly attractive to Lutherans: see her 'Lutheran Uses of Aristotle'. The Protestant churches were faced with a particular need to establish authority, but the breakdown of the institutional authority of the Roman Catholic church affected that church and its civil institutions as well. The contribution of Zabarella to concepts of method may be a particular case in point: see, for instance, Mikkeli, *Aristotelian Response*. Nicholas Jardine has indicated the necessity for further research into the relationship between philosophy and civil life, particularly in the context of the University of Padua and the Venetian republic: see Jardine, 'Keeping Order in the School of Padua'.

[13] For instance, explicit parallels between the study of nature and the study of the Bible can be found in the theology of Jakob Heerbrand, Professor of Theology at the University of Tübingen from 1557 to 1599, and this thinking seems to inform the work of Tübingen's astronomers (although not the natural philosophers): see Methuen, *Kepler's Tübingen*, esp. 107–58.

[14] Harrison, *The Bible, Protestantism, and the Rise of Natural Science*, 78–92.

[15] Harrison, *The Bible, Protestantism, and the Rise of Natural Science*, 90.

adequacy of both ancient and medieval systems of knowledge, leaving the field open for a completely new approach to nature and for the associated development of new theories.

Harrison's discussion of natural history offers a useful interpretative framework for considering the natural philosophy of Philip Melanchthon. There is no doubt that Melanchthon did appeal to the natural world, but, as can be demonstrated from his use of astronomical observation, he did so in the context of a traditional, Aristotelian worldview, and in the expectation that this worldview would be confirmed by precise observations. Although astronomy was traditionally regarded as a mathematical science, and not a part of physics, Melanchthon, in what he understood to be the tradition of Thales of Miletus (but probably drawing not only on Cicero's interpretation of the heavens in *De natura deorum*,[16] but also on Ptolemy's Stoic approach to astronomy[17]), included 'the doctrine of celestial motions and effects' – that is, astronomy and astrology – in his definition of physics.[18] To enable the observer to obtain a better understanding of these motions, Melanchthon recommended the use of the most accurate observations available, and here he particularly praises the measurements made by Copernicus for their contribution to improved accuracy, whether in calculating geographical position, the position of the planets, or the length of the year.[19] However, a comparison of the first (1549) and second (1550) editions of the *Initia doctrinae physicae* suggests that it was for these contributions, and not for Copernicus's reassessment of cosmology or for his heliocentric system, that Melanchthon 'came to love Copernicus more'.[20] Although in the second edition of the *Initia* Melanchthon removed a critical passage suggesting that some astronomers teach the movement of the earth 'either because of their love of novelties or in order to appear clever' and he no longer characterized the renunciation of geocentrism as a game, he also removed any mention of the possibility that Mercury and Venus might orbit the sun rather than the earth, a theory which in 1549 he had discussed but in any case rejected.[21] His tendency in questions of cosmology seems to have been to move away from observational evidence and towards a reaffirmation

---

[16] See, for instance, Hooykaas, *Fact, Faith and Fiction*, 18–19.

[17] Methuen, 'The Role of the Heavens', esp. 402. For the relationship between natural philosophy and ethics in Ptolemy's thought, see Taub, *Ptolemy's Universe*.

[18] Melanchthon, *Initia doctrinae physicae*, CR 13, 183–5, and compare Kusukawa, *The Transformation of Natural Philosophy*, 145–9.

[19] Melanchthon, *De Casparo Crucigero*, CR 11, 839, and compare Melanchthon, *Initia doctrinae physicae*, CR 13, 244 and 262. See also Methuen, *Kepler's Tübingen*, 93 n. 131.

[20] Melanchthon, *De Casparo Crucigero*, CR 11, 839.

[21] See Methuen, *Kepler's Tübingen*, 91–4.

of a Ptolemaic or Aristotelian system.[22] Melanchthon's use of astronomical observation and his application of contemporary work can thus be understood as an attempt to demonstrate the congruence between the prevailing worldview and modern observation, and are thus, to use Harrison's terminology, 'parasitic on prevailing paradigms'. Melanchthon's appeal to the work of Copernicus seems a good example of the humanist approach to the natural world which Harrison describes as common in the sixteenth century.

It would appear, then, that Melanchthon's use of astronomy runs counter to Vivian Nutton's claim that, in his citing of anatomical works, 'Melanchthon's criterion was truth, not Hellenism'.[23] In fact, Nutton's claim seems to be based upon a false dichotomy. It is surely correct, and both Nutton and Kusukawa show convincingly, that Melanchthon believed that a knowledge of anatomy – and thus of the divine order of the natural world – served as a model for ethically good behaviour,[24] just as he believed that observations of the divinely ordained order of the heavens could serve as a model for ordered society. Moreover, it is clear that Melanchthon believed that anatomy could reveal how the soul worked; how spirits, including the Holy Spirit, operated in the body; and how the mind and soul could go wrong.[25] In this sense, Melanchthon was indeed concerned with truth, and it was with this end in mind that he incorporated up-to-date anatomical knowledge into his treatises on the soul, *Commentarius de Anima* (1540) and *Liber de anima* (1552). However, the basis of both these works is Aristotle's *De anima*, to which they offer an extensive, if eclectic, commentary. In his use of anatomy, as in his use of astronomy, Melanchthon's appeal to direct knowledge of the natural world was thus closely related to his humanist rereading of ancient texts.[26] Melanchthon was interested in truth, but, like many of his humanist contemporaries, he regarded truth not as an opponent of, but as a complement to, or even embodied in, Hellenism.

The extent to which this aspect of Melanchthon's approach to natural philosophy influenced Lutheranism is difficult to gauge. Melanchthon's philosophical works were undoubtedly influential and widely read. At

---

[22] Barker notes that in editions of the *Initia doctrinae physicae* printed in Wittenberg in 1562 and in Prato in 1567 the criticisms of Copernicus have been removed or revised (Barker, 'The Role of Religion', 64). These editions, however, were published after Melanchthon's death in 1560, and cannot therefore be cited as evidence for Melanchthon's attitude to Copernicus. The removal of criticisms of Copernicus in the later sixteenth century does, however, seem to indicate that Copernicus's work was increasingly accepted in the context in which these editions were published.
[23] Nutton, 'Wittenberg Anatomy', 16.
[24] Nutton, 'Wittenberg Anatomy', 16–21.
[25] Nutton, 'Wittenberg Anatomy', 20–21.
[26] Frank and Bellucci show how Melanchthon draws upon classical tradition in his articulation of natural philosophy: Frank, *Die theologische Philosophie Philipp Melanchthons*; Frank, 'Gott und Natur'; Bellucci, *Science de la Nature et Réformation*.

a good number of German Lutheran universities, including those in Frankfurt-an-der-Oder, Greifswald, Heidelberg, Königsberg, Rostock, and probably Marburg, natural philosophy was to be taught (at least according to the university statutes) from Melanchthon's *Initia doctrinae physicae*.[27] On the other hand, he was criticized as a crypto-Calvinist for his theological beliefs, and in particular for his teaching of the third use of the law, a doctrine also held by Calvin.[28] In the later sixteenth century the value of his works, including his physics, seems to have been open to doubt in at least one German Lutheran university, namely Tübingen.[29] Moreover, in the later sixteenth century there was a tension within German Lutheranism as to whether natural and moral philosophy should be taught at all.[30] Opponents of natural philosophy probably felt that they were following Luther's recommendation:

> My advice would be that Aristotle's *Physics, Metaphysics, Concerning the Soul* and *Ethics*, which hitherto have been thought to be his best, should be completely discarded together with all the rest of his books that boast about nature, although nothing can be learned from them either about nature or the spirit.[31]

Although Luther does in fact leave an opening for a direct approach to the natural world ('I would almost go as far as to say that a potter understands more about natural things than do these books'[32]), and it can be argued that in general his approach to astronomy and medicine, like Melanchthon's, 'promoted the experimental, experiential aspects of natural sciences',[33] unlike Melanchthon, Luther found it neither necessary nor desirable to incorporate natural philosophy into his intellectual system. Moreover, even when natural philosophy was taught, it is not at all clear that Melanchthon's appeal to the observation of the natural world was integrated into the curriculum. This can be seen from the work of Georg Liebler, Professor for Natural Philosophy at Tübingen from 1552 to 1594. Liebler believed that one law, put in place by God, had shaped both nature and society, and that human intellects should strive to understand this law.

---

[27] Pozzo, 'Die Etablierung des naturwissenschaftlichen Unterrichts', esp. 279–84.
[28] Compare 'Philipp Melanchthon', TRE 22, 392. For the controversies on the meaning of the law between followers of Luther and Melanchthon, see, for instance, 'Gesetz V', TRE 13, 86–7.
[29] Methuen, *Kepler's Tübingen*, 102–3. Kusukawa indicates, however, that in 1536, just after the university had been reformed, Wilhelm Bigot left Tübingen, accused of teaching Aristotle's *De anima* differently from Melanchthon: Kusukawa, 'Lutheran Uses of Aristotle', 170–1. Compare also Kusukawa, 'Natural Philosophy'.
[30] Hofmann, *Artistenfakultät*, 141–2; and compare Methuen, *Kepler's Tübingen*, 61–3.
[31] Luther, *An den christlichen Adel*, WA 6, 457; LW 44, 200.
[32] Luther, *An den christlichen Adel*, WA 6, 458; LW 44, 200.
[33] Maaser, 'Luther und die Naturwissenschaften', 40.

In his view, the natural world retained aspects of the 'original, pristine wisdom which was given to our first parents'; God could thus be known through the contemplation of nature. However, although Liebler was convinced that such 'contemplation' should include an accurate explanation of all the parts of philosophy, he taught that it was through the philosophy of Plato or of Aristotle that the attributes of the divine mind which has created the world may be recognized.[34] In general, the content of natural philosophy at late-sixteenth-century German Lutheran universities remained primarily Aristotelian, with little or no direct reference either to new discoveries or to the natural world.[35] Although Melanchthon's *Initia doctrinae physicae* was used, it seems simply to have supplanted Aristotle's physics as the authority from which a text-based natural philosophy should be taught.[36]

Nonetheless, it is apparent that Melanchthon's redefinition of natural philosophy to take account of the findings of anatomical dissection and of more accurate astronomical observations – that is, his appeal to nature – did serve as an inspiration, for instance, to astronomers and others who were eager to find authority for a new approach to the natural world. The question remains whether this is sufficient for the appeal to the natural world to be categorized as distinctively Lutheran, especially since it would appear from Harrison's analysis that Melanchthon's approach was entirely congruent with the interests of humanist scholarship, which would mean that it was neither exclusively Lutheran nor exclusively Protestant.[37] It is clear that those taking new interest in observing the natural world were drawn from a variety of the increasingly disparate theological camps of the sixteenth century and thus formed a group that transcended confessional boundaries (which were in any case not yet fully established in the late sixteenth century): thus an emphasis on observation can be found in the work of da Vinci as well as of Dürer, of Galileo and Copernicus as well

---

[34] Methuen, *Kepler's Tübingen*, 152–3.

[35] Note too that although Jacob Schegk also referred to Vesalius' results, he used them rather differently than did Melanchthon: Kusukawa, 'Lutheran Uses of Aristotle', 170. This medical knowledge may well have been included in his lectures on dialectics, rather than on natural philosophy: Hofmann suggests that in Tübingen in the 1560s and 1570s, Jakob Schegk's lectures on Aristotle's *Organon* attracted medical students because of the number of medical examples he used: Hofmann, *Artistenfakultät*, 143.

[36] See e.g. Wollgast, *Philosophie in Deutschland*, 147. Wollgast argues that Melanchthon's work became the authority for Lutheran natural philosophy, while within Calvinist circles, the work of Theodore Beza was regarded as authoritative (ibid.). However, Strohm has argued that Beza understood himself as the true inheritor and mediator of the tradition of Melanchthon: Strohm, 'Beobachtungen zur Melanchthon-Rezeption'.

[37] Humanism was of course deeply influential throughout Europe and deeply influenced the development of all three main confessions. For a brief discussion, see, for instance, Rex, 'Humanism'; for more detail compare Nauer, *Humanism and the Culture of Renaissance Europe*. For a discussion of the influence of humanism in Italy, see Buck, 'Der Italienische Humanismus'.

as of Kepler. In the late sixteenth century, both Catholic and Lutheran disputants sought to use appeals to the natural world as conclusive arguments in disputes about the accuracy of Aristotle's natural philosophy or his cosmology. Thus, in a dispute about the possibility of the existence of a vacuum, both sides, Telesio and Patrizzi on the one hand, and the Jesuits of Coimbra on the other, appealed to experience, although they were not referring to actual experiments.[38] In Lutheran circles, discussions of the observations of the nova of 1572 reveal a range of very different ways of integrating what had been observed (a new comet or star beyond the moon) with what Aristotle indicates the physical truth to be (that change – and thus the appearance of comets and other phenomena – can take place only in the sublunar sphere).[39] Seventeenth-century Jesuit astronomers and natural philosophers used the work of Galileo in a very similar way to Melanchthon's appeal to Copernicus.[40] In both the Catholic and the Lutheran context, tensions arose between astronomers and natural philosophers who wished to appeal to observation, and those who preferred not to accept challenges to established philosophical and theological systems: both Galileo and Kepler were criticized for their philosophical (in the sense of cosmological) views as well as for their theological positions.[41] The appeal to observation is not restricted to one confession. On the other hand, the prevalence of Lutheran voices in, for instance, astronomical discussions indicates that much more work needs to be done on the differences between Lutheran, Calvinist and Catholic approaches to the technical and practical arts before this dispute can be decided.

The doctrine of providence

Kusukawa has suggested that Melanchthon's theological justification of his approach to the natural world – namely his belief that the order which might be observed in the natural world revealed divine providence and could serve as a model for the ordering of human society – might be the key for understanding a Lutheran natural philosophy. This classification is problematic, in part because the appeal to God's providence as the ground for the study of the natural world is an important aspect of the thought of a wide range of thinkers, including Galileo and astronomers of the

---

[38] Hooykaas, *Fact, Faith and Fiction*, 201–2. For the move from an appeal to experience to the conducting of experiments, see Dear, *Discipline and Experience*.
[39] See Chapter 4 above.
[40] Hellyer, 'Jesuit Physics', 540–1.
[41] For Galileo, see for instance Bellone, *Galileo Galilei*, and compare McMullin, *Galileo*. For Kepler, see Caspar, *Kepler*, and Hübner, *Die Theologie Johannes Keplers*.

Reformed tradition,[42] but also because Luther's understanding of providence is rather different from that of Melanchthon. In his study of the doctrine of providence, Reinhold Bernhardt argues that, although Luther never worked out his ideas about providence systematically, his understanding of providence focused, not upon the order of the world, but upon God's action in constantly renewing his created universe according to his will. Although Luther recognized the existence of a *providentia generalis*, his protest against any form of *theologia gloriae* led him to animadvert on those who seek to use natural knowledge to understand the things of God: thus, while he conceded the existence of natural knowledge, Luther believed that it could do no more than underline what is already known through revelation.[43] Therefore, the suggestion that 'a Lutheran *must* be able to achieve knowledge of God's providential plan'[44] is a misinterpretation of Luther's position (and, indeed, comes close to precisely the kind of theological overconfidence which he abhorred).

Kusukawa bases her claim that Melanchthon's doctrine of providence is Lutheran upon her conviction that his theology of providence was fundamentally different from – and opposed to – that of Zwingli.[45] She suggests that Zwingli's theology 'was centred on the absolute Providence of God', in such a way that 'the divinity was understood in opposition to creatures'. In contrast, Luther, and, she implies, Melanchthon, 'saw God's divinity in precisely that creatureliness'.[46] Zwingli, she believes, conceives God as sovereign over – and thus, she implies, distanced from – the world, while Luther and Melanchthon believe God to be found – or perhaps better, to be recognizable – in the natural world. Leaving aside the problems with her reading of Luther, which Bernhardt's work suggests is flawed, Kusukawa's interpretation of Zwingli seems to be based upon a misapprehension. Zwingli's understanding of God as 'sovereign', in the sense of the director of every event, served not to distance God from the world, but to ensure God's close involvement in every event. In his sermon *De providentia*, Zwingli identifies God with every active or guiding principle which has been named by the philosophers as such; by doing so, he 'attribute[s] to God what others have attributed to nature',[47] and identifies God's providential action closely both with the natural world as created by God and with

---

[42] Hooykaas suggests that the doctrines of creation and providence as defined in the Calvinist *Confessio Belgica* could and did serve this purpose: see *Religion and the Rise of Modern Science*, 105–6 (although his argument draws also on the work of Kepler, who was at least technically Lutheran).
[43] Bernhardt, *Was heißt 'Handeln Gottes'?*, 61–86, esp. 84–5. See also Chapter 1 above.
[44] Barker, 'The Role of Religion', 61 (author's italics).
[45] Kusukawa, *The Transformation of Natural Philosophy*, 161–3.
[46] Kusukawa, *The Transformation of Natural Philosophy*, 163.
[47] Stephens, *The Theology of Huldrych Zwingli*, 86.

philosophical explanations of it. Zwingli's perception of the relationship between providence and natural philosophy thus seems to differ from that held by Melanchthon less radically than Kusukawa suggests. There were indeed differences: Melanchthon believed the natural world to possess a fundamental order, imposed upon it by God, revelatory of God's purpose, and accessible, at least to some extent, through natural philosophy. Zwingli, on the other hand, focused not on an underlying order but on God's power to direct every event and every action. However, for him, as for Melanchthon, this power is revealed, not only in the Holy Scriptures, but in the way that philosophers have described the world.[48] Additionally, and in strong contrast to Luther, both Melanchthon and Zwingli were happy to draw upon the insights of secular authors to assist their understanding of God's providence as revealed in the natural world.

The problem of using the doctrine of providence as a confessional marker is compounded by Bernhardt's observation that Calvin's doctrine of providence, and particularly its focus on God's action, was closer to Luther's than to Melanchthon's, so that 'the first systematic presentation of Lutheran belief in Providence is found in the work of Calvin'.[49] Moreover, although Bernhardt believes that it was Melanchthon's doctrine of providence, and his focus on natural knowledge, which formed the foundation of the understanding of providence in the Lutheran orthodoxy of the seventeenth century,[50] he also argues that this understanding was closely related to that of Calvinist orthodoxy.[51] It is thus possible to trace the influence of both Luther and Melanchthon on doctrines of providence held within both Calvinism and Lutheranism. This result is not really surprising, given that, as Christoph Strohm has shown, a number of significant late-sixteenth-century Calvinists (notably Theodore Beza) understood themselves to be the true inheritors and disciples of Melanchthon.[52] Kusukawa has herself indicated that Melanchthon's doctrine of the soul attracted the interest of Calvinist as well as Lutheran scholars; his approach, together with its underlying theology, seems to have appealed to both.[53]

---

[48] See also Chapter 1 above.
[49] Bernhardt, *Was heißt 'Handeln Gottes'?*, 88–9 (quote on 88).
[50] Bernhardt, *Was heißt 'Handeln Gottes'?*, 123–43, esp. 126 and 131–3.
[51] Bernhardt, *Was heißt 'Handeln Gottes'?*, 144–7.
[52] Strohm, 'Beobachtungen zur Melanchthon-Rezeption'. Lohr has suggested that natural philosophy occupies a different place in the intellectual systems of Catholic, Lutheran and Calvinist thought, especially with regard to defining the distinctions and overlaps between natural philosophy and the subject matter and methodology of metaphysics. He notes that it was Calvinist, rather than Lutheran, Aristotelianism (or perhaps better, scholasticism) which moved towards a system of knowledge: see 'Metaphysics and Natural Philosophy as Sciences', esp. 292–3. However, as Strohm demonstrates, Calvinist theories of systematization, and in particular their use of *loci communes*, were probably derived from Melanchthon's work.
[53] Kusukawa, 'Natural Philosophy', 450–2.

The difference of focus in the approach taken by Melanchthon from that taken by Zwingli can be formulated in terms of a different focus within their respective doctrines of providence. Melanchthon, with his sense of order, focuses more (but not exclusively) on general providence, while Zwingli might be said to be more concerned to emphasize God's actions in the form of special providence. Although, as I have argued elsewhere, both these approaches were used in the sixteenth century by astronomers keen to explain the observations they had made, and although both could aid observers in their criticism of Aristotle,[54] a focus on special providence could lead away from an interest in the natural world, as it did for Luther and Calvin.[55] In contrast, Melanchthon's assumption that a fundamental order exists and can be observed in the natural world is more conducive to the search for laws of nature associated with the development of modern science, and can be identified as an emphasis on what in the seventeenth century would come to be identified more clearly as natural theology. It is, however, by no means clear that this assumption is specifically Lutheran: the sense of order which permeates Melanchthon's theological and philosophical system also shapes the thought of the Calvinist Lambert Danaeus, who, like Melanchthon, understands natural philosophy both to inform ethics and to aid the understanding of God's creative work.[56] But that sense of order is found too in the work of the Jesuit Francesco Suarez; indeed, Bernhardt suggests that the particular form of *providentia generalis* found in Lutheran orthodoxy, and in particular its association with the theory of accommodation (used by the Lutheran astronomer Christoph Rothmann in the later sixteenth century to justify discrepancies between his observations and Scripture[57]), can be traced back, not to Melanchthon, but to Suarez.[58]

The appeal to the notion of providence, in the sense of an emphasis upon general providence, and thus on natural theology, can thus be found, not only in Lutheran natural philosophy, but in both Calvinist and Catholic natural philosophy as well. Charles Lohr argues that in the course of their search for the common principles of all sciences, Suarez, together with his fellow Jesuit Perera and the Dominicans Cajetan and Javelli, 'constructed a new science of metaphysics based on the revealed idea of creation, transforming Aristotle's philosophy into the higher science' (i.e. metaphysics),[59] while within Calvinism 'natural theology formed an integral part of the

---

[54] See Chapters 4 and 5 above.
[55] For this effect in the work of Jakob Andreae and Matthias Haffenreffer, see Methuen, *Kepler's Tübingen*, 116–17, 125–8 and 151–2; compare also Chapter 5 above.
[56] Strohm, *Ethik im frühen Calvinismus*, 650.
[57] See Granada, 'Il Problema Astronomico-cosmologico'.
[58] Bernhardt, *Was heißt 'Handeln Gottes'?*, 127, 153–4.
[59] Lohr, 'Metaphysics and Natural Philosophy as Sciences', 287.

cognitio Dei perfecta at which theology aimed'.[60] It would seem, then, that an emphasis on the doctrine of providence, and particularly on general providence, far from being specifically Lutheran, might be a feature common to neo-scholasticism in all its manifestations in the late sixteenth and early seventeenth centuries. More work needs to be done to establish whether this is indeed so, and to investigate whether, as seems probable, this doctrine can be associated with a Christian anthropology which affirms a God-given ability to recognize this order, either through 'natural light' or a similar divinely given reflection of God's mind, which is thus innate, albeit damaged by the Fall.[61]

Ubiquity and the theology of the Eucharist

It is, therefore, necessary to look elsewhere if adequate criteria are to be found for a specifically Lutheran approach to the natural world. Given that the theology of the Eucharist played a central role in defining confessional difference during the sixteenth century, differences in the teaching of natural philosophy arising from the confessional differences in defining the doctrine of the Eucharist would seem a fruitful possibility for identifying a distinctively Lutheran natural philosophy, especially since the different doctrines of the Eucharist involved different conceptions of the properties of matter, or, more specifically, of the location of Christ's physical body.[62]

Eucharistic controversy did not, of course, begin in the sixteenth century. The medieval church had come to teach the doctrine of transubstantiation, as articulated by the Fourth Lateran Council of 1215, refined by Thomas Aquinas, and finally stated at the Council of Trent. The establishment of this doctrine was a long process, and it caused grave difficulties for natural philosophy.[63] Advocates of transubstantiation taught that through

---

[60] Lohr, 'Metaphysics and Natural Philosophy as Sciences', 291. Lohr believes that of the three confessions present in Germany it was Lutheranism that was most reluctant to admit 'the necessity of an independent natural theology' (ibid.). However, there were strands within Calvinism which preferred to emphasize the unpredictability of the natural world and to emphasize Biblical revelation. See Chapter 1 above.

[61] The doctrine of 'natural light', found in Melanchthon's thought, is associated with his doctrine of providence and shapes his understanding of how the natural world may be known; for a detailed discussion, see Bellucci, *Science de la Nature et Réformation*. The theology of natural light is also associated with the understanding that the natural world functions according to a divine law in the theology of Thomas Aquinas: see, for example, *Summa Theologica*, 1a2ae, 109, 1 (ST 30, 67–73).

[62] Ann-Charlott Trepp points to another possibility: that nature mysticism as exemplified in the work of Johann Arndt might shed light on these questions. See ' "Nature" and Religious Practice'.

[63] For a detailed discussion of the complexities of reconciling the doctrine of transubstantiation with Aristotelian natural philosophy in the Middle Ages, see Bakker, *La raison et le miracle*.

the words of consecration, the substance of Christ's body and blood became physically and irrevocably present in the bread and wine, replacing the substance of the bread and wine, while the accidents of the bread and wine remained those of bread and wine.

Associated with the doctrine of transubstantiation were a range of liturgical practices which centred on the adoration of the consecrated host, but which could quickly slip into superstition or magic.[64] This was one reason why the doctrine was severely criticized by the Reformers. However, the Reformers themselves held radically different understandings of the Eucharist, ultimately resulting in irreconcilable differences in their eucharistic doctrines. Thus, while Zwingli criticized not only the doctrine of transubstantiation but also the idea of a real physical presence, understanding the Eucharist to be a memorial meal,[65] Luther held to the understanding that Christ was physically present in the Eucharist, but rejected the doctrine of transubstantiation as an unnecessary philosophical explanation of what God intended to be a divine mystery.[66] Pushed by Zwingli and his followers to explain how Christ's body could be physically present, Luther drew upon the doctrine of the ubiquity of Christ, teaching that through his divine person, Christ's physical body partakes of the divine attribute of omnipresence, and can thus be present everywhere and in every Eucharist.[67] Finally, Calvin taught that Christ was present in the Eucharist as a real, but spiritual, presence.

Amos Funkenstein understands the rejection of the doctrine of transubstantiation not only to bring about, but to result from, a different understanding of nature. More specifically, he believes that the doctrine of consubstantiation (which he ascribes to Luther) may have been informed by a view of nature as both uniform and animated (in the sense of driven by the soul) which arose in late-medieval nominalism and developed further during the Renaissance.[68] Against this background, he suggests, consubstantiation was the more attractive doctrine: 'Imagining interpenetrating substances was, to the Middle Ages, no less a conceptual problem than conceiving accidents without their proper subject. To a more Stoic-oriented sense of nature, the complete interpenetration of bodies, *velut ferrum ignitum*, became much less repugnant to common sense.'[69] This thesis does not

---

[64] See, for instance, Scribner, 'Ritual and Popular Religion', 35–6.

[65] For Zwingli's theology of the Eucharist, see Stephens, *The Theology of Huldrych Zwingli*, 218–59, and compare Lohse, 'Dogma und Bekenntnis in der Reformation', 51–5.

[66] Luther regarded the doctrine of transubstantiation as the second 'captivity' of the Eucharist: see Luther, *De captivitate*, WA 6, 508–12; LW 36, 28–35.

[67] The term 'consubstantiation', often associated with Luther's doctrine of the Eucharist, was never used by him, but may be derived from the formulation in article X of the *Confessio Augustana variata* of 1540.

[68] Funkenstein, *Theology and the Scientific Imagination*, 57–71.

[69] Funkenstein, *Theology and the Scientific Imagination*, 71.

sit well with the thought of Luther, but it might offer a useful insight into Melanchthon's interests. The arguments cited during the eucharistic controversies between Luther and Zwingli centred upon issues of Biblical interpretation, that is, of the meaning of language in general and of scriptural language in particular,[70] of questions of dialectics and of rhetorical figures. Arguments from natural philosophy occurred only seldom, precisely because the proponents of the new theology were seeking to cleanse theology of philosophical arguments. Under pressure to justify his position, Luther appealed to the doctrine of ubiquity in an explicit attempt to avoid the use of philosophical terminology and arguments in explaining this divine mystery.[71] Luther's rejection of the doctrine of transubstantiation does not presuppose a rejection of the natural-philosophical explanation of matter which forms its basis, but rather a rejection of the assumption that one particular argument of natural philosophy might be declared an infallible explanation for a divine mystery, the explication of which should, in his eyes, be a purely theological matter. The emphasis of Luther's arguments lay in his consistent attempts to exclude philosophical speculation from the study and discussion of theology.[72] It is thus difficult to associate Luther's theology of the Eucharist directly with one particular view of nature or approach to natural philosophy.

Melanchthon's philosophical approach, unlike Luther's, does fit with Funkenstein's suggestion that the doctrine of consubstantiation might appeal to a believer in a Stoic approach to nature, and indeed it was Melanchthon who introduced the idea that Christ's body was present with (*con*) the bread in the Eucharist, and who thus enabled the coining of the term 'consubstantiation'. However, although Melanchthon's understanding that natural philosophy may provide a model for moral philosophy is almost certainly Stoic in origin, it is not, as we have seen, exclusively Lutheran. Barker suggests that it is the doctrine of Christ's physical presence in the Eucharist, explained in terms of Christ's ubiquity, which forms 'the basis for the Lutheran belief in the universal presence of a providential deity, whose design or plan may be known through the study of nature'.[73] This seems not to take account of the differences amongst Lutherans, all of whom believed in the physical presence of Christ's body in the Eucharist, but not all of whom valued the study of the natural world. Nonetheless, amongst natural philosophers, theological convictions about the way in

---

[70] For a summary of the conflict between Luther and Zwingli, see Lohse, 'Dogma und Bekenntnis in der Reformation', 46–64, and compare also Stephens, *The Theology of Huldrych Zwingli*, 235–50.
[71] Luther, *De captivitate*, WA 6, 509–11; LW 36, 31–4.
[72] For Luther's use of Aristotle, see Dieter, *Der junge Luther und Aristoteles*.
[73] Barker, 'The Role of Religion', 62.

which Christ was present in the Eucharist, and the consequent necessity of distinguishing between different ways of 'being present',[74] gave rise to the associated philosophical question of what it meant to speak of 'presence' in the first place. The discussions of eucharistic theology do seem to have had implications for the way in which nature, and particularly matter, could be conceived.

The Catholic doctrine of transubstantiation, with its terminology of substance and accidents, assumed a theory of matter which not only allowed these terms to have meaning but also allowed a body (Christ) to be present by taking on accidents which were not natural to it.[75] The doctrine of transubstantiation has been shown to have shaped what could officially be taught by Jesuit natural philosophers;[76] can the same be said of Lutheran doctrines of the universe? Certainly, the real physical presence of Christ in the Eucharist, whether taught by Roman Catholics or by Lutherans, required that the physical body of Christ be present in more than one place at any one time. Natural philosophy conducted in the context of this doctrine thus needed concepts of space and place which allowed natural bodies (such as the body of Christ) not to be circumscribed by a finite, definite place.[77] Both the doctrine of transubstantiation and Lutheran understandings of the real physical presence, therefore, required the truth of certain presuppositions about the nature of matter or of location. This, however, was not the case for the Calvinist theology of the Eucharist, which only required Christ to be present spiritually.[78] The doctrine of ubiquity and the consequences of this doctrine for the location of Christ's body may indeed have bound Lutheran natural philosophers in their discussions of space and place in a way which was not true for their Calvinist counterparts.[79] This affected the way that natural philosophers were able to accept Descartes' philosophy of matter, and could indicate a precise, technical confessional difference within natural philosophy.

---

[74] See, for instance, Lohse, 'Dogma und Bekenntnis in der Reformation', 57.

[75] This despite the fact that the Council of Trent did not intend eucharistic doctrine to be dependent on a specific theory of matter: see Artigas *et al.*, 'New Light on the Galileo Affair'.

[76] See, for instance, Hellyer, ' "Because the Authority of my Superiors Commands" ', esp. 327–8.

[77] See Leijenhorst, 'Place, Space and Matter in Calvinist Physics', 524.

[78] Leijenhorst investigates the concepts of space as used by four Calvinist natural philosophers: Petrus Ramus, Bartolomäus Keckermann, Clemens Timpler and Johann Heinrich Alsted. All four reject any possibility of multi-location and require that a natural body must occupy a particular *localitas*: see Leijenhorst, 'Place, Space and Matter in Calvinist Physics'. If this observation can be generalized to all Calvinist natural philosophers, it may indicate a contributory factor to the reception of Descartes' philosophy in the (Calvinist) Netherlands: see Howell, 'Science, Religion and the Common Good'.

[79] Compare Leijenhorst and Lüthy, 'The Erosion of Aristotelianism'.

Lutheranism and natural philosophy

Defining Lutheran natural philosophy, or a Lutheran approach to the natural world, is not as easy as it might seem at first sight. Not everyone who was a Lutheran in the sense of being a follower or associate of Luther would have been defined as Lutheran in the later, confessional sense. This is particularly true of Melanchthon, a key player in this debate. While there can be no doubt that Melanchthon had a keen theoretical and theological interest in the study of the natural world, and that he imparted his interest to a circle of his students, who in turn passed it on, it is not entirely clear that this interest was Lutheran in a confessional sense. That it met fruitful ground is due at least in part to the political structures of the German empire, for its lack of centralized structures and its relatively high numbers of courts and of 'national' universities enabled a culture in which at least some of those who wished to carry these ideas further were able to find opportunities to do so.[80] Within German Lutheranism, and perhaps particularly within the circles of those (often astronomers) who addressed the study of nature from the interface between university and court life,[81] Melanchthon's work and ideas provided those who wished to observe nature (and perhaps particularly the heavens) accurately with an authority which could be used to maintain the truth of what they observed, particularly when their observations contradicted the established 'truths' of natural philosophy. Within Lutheran circles, the care taken by astronomers to interpret Scripture in such a way as to allow their observations to be accepted suggests that they were particularly concerned with the status of Scripture and took its authority seriously. It may therefore be true to say that the political context of German Lutheranism, particularly for those students of Melanchthon who took seriously his strictures regarding the necessity of studying the natural world, was particularly open to the development of a body of (particularly astronomical) knowledge which, although not immediately associated with natural philosophy, came to be used in to critique and further develop it.

---

[80] See, for instance, Moran, 'Wilhelm IV of Hesse-Kassel', and the papers collected in Moran (ed.), *Patronage and Institutions*. The lack of a centralized government in the German empire meant that, although individual German states imposed censorship, it was impossible for a particular methodology to be banned wholesale, as it was, for instance, in Spain: García-Ballester, 'Inquisition and Minority Medical Practitioners', 164.

[81] For the role of astronomers in German courts and states during the sixteenth century, see Schöner, *Mathematik und Astronomie an der Universität Ingolstadt*, who focuses on Catholic Ingolstadt, but derives conclusions which are congruent which those Robert Westman has reached and which can thus be extended to the Lutheran states: compare Westman, 'The Astronomer's Role', and idem, 'Humanism and Scientific Roles'.

There are still many avenues to be explored concerning the way in which the content of natural philosophy was influenced by theological doctrines (and vice versa). On the basis of current knowledge, the tradition of Philip Melanchthon can be seen to have influenced both the development of a neo-scholastic, Aristotelian-based natural philosophy (which may later have split into two schools which did have distinctive Lutheran and Calvinist features in their teachings about matter, space and time), and a more observation-based astronomical trend, which would ultimately offer fundamental criticisms of central Aristotelian themes. It might be argued that Melanchthon's thought was a significant factor for the development, not so much of a Lutheran natural philosophy, but of two strands of thought which were important to the development of early modern intellectual life.

Bibliography

Abbreviations

CO   *Ioannis Calvini Opera*, ed. Guilielmus Baum, Eduardus Cunitz and Eduardus Reuss (Braunschweig: C. A. Schwetske and M. Bruhm) = CR 29–87
CR   *Corpus Reformatorum*, ed. Karl Gottlieb Bretschneider (Berlin: Schwetschke; Leipzig: Heinsius; Zurich: Theologischer Verlag)
ET   English translation
KGW   Johannes Kepler, *Gesammelte Werke* (Munich: Beck)
LCC   Library of Christian Classics (see entries under individual authors for bibliographical details of particular volumes)
LW   *Luther's Works*, ed. Jaroslav Pelikan (St Louis: Concordia)
NRSV   New Revised Standard Version
ST   Thomas Aquinas, *Summa theologiae*; ET ed. Thomas Gilby (London: Blackfriars with Eyre & Spottiswoode, 1964–81)
TRE   *Theologische Realenzyklopädie* (Berlin: Walter de Gruyter, 1976–2004)
WA   *D. Martin Luthers Werke Kritische Gesamtausgabe* (Weimar: Hermann Böhlaus Nachfolger)

Manuscripts and archival material

Correspondence on the 1572 nova held in the Hauptstaatsarchiv, Stuttgart (HStAS), A274 Bü 21, including:

    Anonymous, 'Iudicium de his qui misere in Gallijs trucidati sunt, & de noua stella' (1573).
    Eisenmenger, Samuel, 'De Stella Nova apparente in stellae coelo consideration et obseruatio' (1573).
    Maestlin, Michael, 'Noua stella' (n.d., prob. 1572).

Correspondence regarding Nicodemus Frischlin, HStAS, A274 Bü 45, Bü 46, Bü 47, esp. Bü 46, ## 19, 26, 28, concerning his astronomy textbook, including:

    Maestlin, Michael, 'Iudicium M. Moèstlini de opere Astronomico D. Frischlini' (1586), Bü 45, #30.

Correspondence between Duke Ludwig and Michael Maestlin, HStAS, A202, Bü 2551.

Printed sources

Andreae, Jacob, *Christliche, notwendige vnd ernstliche Erinnerung Nach dem Lauff der irdischen Planeten gestelt, Darauß ein jeder einfeltiger Christ zusehen, was für glück oder vnglück Teutschland diser zeit zugewarten. Auß der vermanung Christi Luc 21 in fünff Predigen verfasset* (Tübingen: n.p., 1567).

Anonymous, *Herrlich Bedencken Des tewren Mannes Gottes Lutheri seligen vom dem itzundt newen Beptischen Calender* . . . (no place: no publisher, 1583).

Aquinas, *Summa Contra Gentiles*.

Brahe, Tycho, *Astronomiae Instauratae Progymnasmatum*, in *Tychonis Brahe Danis Opera Omnia*, vols 2–3, ed. J. L. E. Dreyer (Copenhagen: Libraria Gyldendaliana, 1913–16).

Busch, Georgius, *Von dem Cometen, welcher in diesem 1572. Jar in dem Monat Novembris erschienen* (Erfurt: n.p., 1572).

Calvin, John, *Commentarius in Genesin*, CR 51 (*Calvini Opera Omnia* 23), 13–622; ET *Genesis*, tr. and ed. Alister McGrath and J. I. Packer (Wheaton, Ill. and Nottingham: Crossway, 2001).

Calvin, John, *Institutio christianae religionis* (edns of 1536 and 1559), CR 29–30 (CO 1–2); ET *Institutes of the Christian Religion*, tr. Ford Lewis Battles, ed. John T. McNeill, Library of Christian Classics 20, 21 (Louisville, Ky: Westminster John Knox Press, 1960).

Calvin, John, *Iohannis Calvini Commentarius in epistolam Pauli ad Romanos*, ed. T. H. L. Parker (Leiden: Brill, 1981). ET *The Epistles of Paul the Apostle to the Romans and to the Thessalonians*, tr. Ross Mackenzie, ed. David W. Torrance and Thomas F. Torrance (Edinburgh: Oliver and Boyd, 1961).

Calvin, John, *Responsi ad Sadoleti Epistolam*, CR 33 = CO 5, 385–416; ET in *Calvin: Theological Treatises*, tr. and ed. John K. S. Reid, Library of Christian Classics 22 (London: SCM Press, 1954), 221–56.

Clavius, Christopher, *Josephi Scaligeri Enlenchus et castigio Calendarij Gregoriani* (Rome: Zunnetus, 1595).

Crusius, Martin, *Corona Anni: Hoc est explicatio euangeliorum et epistolarum quae diebus festis proponuntur* (Wittenberg: Lautentius Seuberlich, 1603).

Frischlin, Nicodemus, *Consideratio nouae stellae, quae mense Nouembri, anno salutis MDLXXII . . . apparuit* (Tübingen: Georg Gruppenbach, 1573).

Graminaeus, Theodorus, *Erklerung oder Auslegung eines Cometen, so nuhn ein gutte Zeit von Martini des nechst vergangenen Jars bis auff den dritten Februarii diese jetzt lauffenden MDLXXIII Jars am Himmel vernommen und noch bey nächtlicher Zeit gesehen wird* (Cologne: n.p., 1573).

Hafenreffer, Matthias, *Disputatio de providentia Dei* (Tübingen: n.p., 1602).

Hafenreffer, Mathias, *Loci theologici* (Tübingen: Georg Gruppenbach, 1600).

Heerbrand, Jakob, *Compendium theologiae* (Leipzig: Georg Gruppenbach, 1573).

Heerbrand, Jakob, *De adiaphoris, et Calendario Gregoriano* (Tübingen: Alexandrus Hockius, 1583).

Heerbrand, Jakob, *De scripturae sacrae interpretatione* (Tübingen: Alexandrus Hoggius [sic], 1582).

Heerbrand, Jakob, *Disputatio de Magia ex cap. 7. Exo.* (Tübingen: Ulrich Morhart d.Ä., 1570).

Heerbrand, Jakob, *Ein Predig von dem erschrockenlichen Wunderzeichen am Himel, dem newen Cometen oder Pfawenschwantz* (Tübingen: Georg Gruppenbach, 1577).

Kepler, Johannes, *Astronomia Nova*, KGW 3; ET *Astronomia Nova (The New Astronomy)*, tr. William H. Donahue (Cambridge: Cambridge University Press, 1992).

Kepler, Johannes, *Harmonices Mundi*, KGW 6; ET *The Harmony of the World*, tr. E. J. Aiton, J. M. Duncan and J. V. Field (Philadelphia: American Philosophical Society, 1997).

Kepler, Johannes, *Mysterium Cosmographicum*, 1st edn (1596) in KGW 1, 2nd edn (1622) in KGW 8; ET *Mysterium Cosmographicum (The Secret of the Universe)*, tr. A. M. Duncan, introd. E. J. Aiton (Norwalk, Conn.: Abaris Books, 1999).

Leovitus, Cyprian, *Von dem neuen Stern: Bericht Cypriani von Leovitz Mathematici zu Laugingen von dem newen Stern oder Cometen, welcher gesehen ist wordem im November vnd December des 1572. auch im Januario vnd Februario des 1573. Jars* (Laugingen: n.p., 1573).

Leovitus, Cyprian, *De nova stella: Iudicium Cypriani Leovitii a Leonicia, Mathematici, de stella nova sive Cometa, viso mense Novembri ac Decembri, Anni Domini 1572. Item mense Ianuario & Februario, Anni Domini 1573* (Laugingen: n.p., 1573).

Locke, John, *An Essay Concerning Human Understanding*, ed. Peter H. Nidditch (Oxford: Oxford University Press, 1975).

Luther, Martin, *An den christlichen Adel deutscher Nation von des christlichen Standes Besserung* (1520), WA 6, 404–69; ET *To the Christian Nobility of the German Nation*, tr. Charles M. Jacobs, revd. James Atkinson, LW 44, 115–217.

Luther, Martin, *De captivitate babylonica ecclesiae praeludium* (1520), WA 6, 484–573; ET *The Babylonian Captivity of the Church*, tr. A. T. W. Steinhäuser, revd Frederick C. Ahrens and Abdel Ross Wentze, LW 36, 5–126.

Luther, Martin, *Diui Pauli apostoli ad Romanos epistola* (1515–16), WA 56, 3–528; ET *Lectures on Romans*, ed. and tr. Hilton L. Oswald, LW 25, 1–524.

Luther, Martin, *Ermahnung zum Frieden auf die zwölf Artikeln der Bauerschaft in Schwaben* (1525), WA 18, 291–334; ET *Admonition to Peace: A Reply to the Twelve Articles of the Peasants in Swabia*, tr. Charles M. Jacobs, revd. Robert Schulz, LW 46, 17–43.

Luther, Martin, *Galatervorlesung* (1531), WA 40/1, 1–688, WA 40/2, 1–184; ET *Lectures on Galatians*, tr. Jaroslav Pelikan, LW 26, 1–421, LW 27, 1–149.

Luther, Martin, *In Genesin Enarrationum* (1535–45), WA 42, 1–683; ET *Lectures on Genesis*, tr. George V. Schick, LW 1, 1–359.

Luther, Martin, *Von den Konziliis und Kirchen* (1539), WA 50, 488–653; ET *On the Councils and Churches*, tr. Charles M. Jacobs, revd Eric W. Griffith, LW 41, 3–178.

Luther, Martin, *Von Weltlicher Uberkeytt, wie weytt man yhr gehorsam schuldig sei* (1523), WA 11, 245–80; ET *Temporal Authority: To What Extent it Should be Obeyed* (1523), tr. J. J. Schindel, revd. Walther T. Brandt, LW 45, 75–131.

Maestlin, Michael, *Alterum Examen novi Pontficialis Gregoriani Kalendarii* (Tübingen: Georg Gruppenbach, 1583).

Maestlin, Michael, *Ausführlicher und Grundtlicher Bericht von der allgemainen und nunmehr bey sechtzehen Hundert Jaren von dem ersten Keyser Julio biß auf jetztige unser Zeit im gantzen H. Römischen Reich gebrachter Jarrechnung oder Kalender* (Heidelberg: n.p., 1583).

Maestlin, Michael, *Consideratio et observatio cometae aetheri astronomica, qui anno MDLXXX . . . apparuit* (Heidelberg: n.p., 1581).

Maestlin, Michael, *Demonstratio astronomica loci stellae novae*, in Frischlin, *Consideratio nouae stellae*, 27–32; repr. in Brahe, *Tychonis Brahe Danis Opera Omnia* 3, 58–62.

Maestlin, Michael, *Observatio et demonstratio cometae aetherae qui anno 1577 et 1578 constitutus in sphaera Veneris apparuit* (Tübingen: Georg Gruppenbach, 1578).

Melanchthon, Philip, *Argumentum et scholia in Officia Ciceronis* (1525), CR 16, 627–80.

Melanchthon, Philip, *Commentarius in Genesin* (1523), CR 13, 761–92.

Melanchthon, Philip, *De astronomia et geographia* (1536?), CR 11, 292–8.

Melanchthon, Philip, *De Casparo Crucigero* (1549), CR 11, 833–41.

Melanchthon, Philip, *De dignitate astrologiae* (1535), CR 11, 261–6.

Melanchthon, Philip, *De dignitate legum* (1543), CR 11, 630–6.

Melanchthon, Philip, *De discrimine evangelii et philosophiae* (n.d.), CR 12, 689–91.

Melanchthon, Philip, *De doctrina physica* (1550), CR 11, 932–9.

Melanchthon, Philip, *De legibus* (1550), CR 11, 908–16.

Melanchthon, Philip, *Enarratio aliquot librorum Ethicorum Aristotelis* (1529), CR 16, 277–416.

Melanchthon, Philip, *Erotemata dialectices* (1547), CR 13, 509–751.

Melanchthon, Philip, *Initia doctrinae physicae* (1549), CR 13, 180–412.

Melanchthon, Philip, *Loci communes 1521 (Lateinsch/Deutsch)*, tr. Horst Georg Pöhlmann (Gütersloh: Gütersloher Verlagshaus, 1993); Latin text also in CR 21, 81–230.

Melanchthon, Philip, *Loci communes* (1535), CR 21, 253–538.

Melanchthon, Philip, *Loci communes* (1543), CR 21, 601–1106.

Melanchthon, Philip, *Philosophiae moralis epitomes* (1538), CR 16, 21–164.

Melanchthon, Philip, *Praefatio in arithmeticen* (1536), CR 11, 284–92.

Melanchthon, Philip, *Praefatio in libros de iudiciis navitatum Iohannis Schoneri* (1558?), CR 5, 817–24.

Melanchthon, Philip, *Praefatio in theoricae novae planetarum* (1535), CR 2, 814–21.

Scaliger, Joseph, *De emendatione temporum* (Paris: Nivellius, 1583).

Zwingli, Huldrych, *Farrago annotationum in Genesim* (1527), CR 100, 1–290.

Zwingli, Huldrych, *Sermonis de providentia Dei anamnena* (1530), CR 93/3, 1–230; German tr. *Die Vorsehung*, in Huldrych Zwingli, *Schriften*, 4 (Zurich: Theologischer Verlag, 1995), 133–279.

Secondary literature

Allen, Diogenes, *Philosophy for Understanding Theology* (London: SCM Press, 1985).
Ariew, Roger, 'Theory of Comets at Paris during the Seventeenth Century', *Journal of the History of Ideas* 53 (1992), 355–72.
Artigas, Mariano, Martinez, Rafael, and Shea, William, 'New Light on the Galileo Affair', in Brooke and Ihsanoglu (eds), *Religious Values and the Rise of Science in Europe*, 145–66.
Atkinson, Catherine, *Inventing Inventors in Renaissance Europe: Polydore Vergil's* De inventoribus rerum, Spätmittelalter und Reformation Neue Reihe 33 (Tübingen: Mohr Siebeck, 2007).
Bakker, Paul J. J. M., 'La raison et le miracle. Les doctrines eucharistiques (c. 1250–c. 1400): Contribution à l'étude des rapports entre philosophie et théologie', unpub. doctoral thesis, Catholic Univerity of Nijmegen, 1999.
Baldini, Ugo, 'Christoph Clavius and the Scientific Scene in Rome', in Coyne et al. (eds), *Gregorian Reform of the Calendar*, 137–69.
Barker, Peter, 'Copernicus, the Orbs, and the Equant', *Synthese* 83 (1990), 317–23.
Barker, Peter, 'The Optical Theory of Comets from Apian to Kepler', *Physis: Rivista Internazionale di Storia della Scienza*, n.s. 30 (1993), 1–25.
Barker, Peter, 'The Role of Religion in the Lutheran Response to Copernicus', in Margaret J. Osler (ed.), *Rethinking the Scientific Revolution* (Cambridge: Cambridge University Press, 2000), 60–88.
Barker, Peter and Goldstein, Bernard R., 'The Role of Comets in the Copernican Revolution', *Studies in History and Philosophy of Science* 19 (1988), 299–319.
Baur, Jörg, 'Luther und die Philosophie', *Neue Zeitschrift für systematische Theologie* 26 (1984), 13–28.
Bellone, Enrico, *Galileo Galilei: Leben und Werk eines unruhigen Geistes*, 2nd edn (Heidelberg: Spektrum der Wissenschaft 2002).
Bellucci, Dino, 'Gott als mens: Die "physica aliqua definitio" Gottes bei Philipp Melanchthon', in Rhein and Frank (eds), *Melanchthon und die Naturwissenschaften seiner Zeit*, 59–71.
Bellucci, Dino, *Science de la Nature et Réformation: La physique au service de la Réforme dans l'enseignement de Philippe Mélanchthon* (Rome: Edizioni Vivere, 1998).
Benin, Stephen D., *The Footprints of God: Divine Accommodation in Jewish and Christian Thought* (Albany, NY: State University of New York Press, 1993).
Bernhardt, Reinhold, *Was heißt 'Handeln Gottes'? Eine Rekonstruktion der Lehre von der Vorsehung* (Gütersloh: Gütersloher Verlagshaus, 1999).
Blackwell, Constance and Kusukawa, Sachiko (eds), *Philosophy in the Sixteenth and Seventeenth Centuries: Conversations with Aristotle* (Aldershot: Ashgate, 1999).
Blumenberg, Hans, *The Genesis of the Copernican World* (Cambridge, Mass.: MIT Press, 1987).

Bouwsma, William J., *John Calvin: A Sixteenth Century Portrait* (Oxford and New York: Oxford University Press, 1988).
Brooke, John Hedley, *Science and Religion: Some Historical Perspectives* (Cambridge: Cambridge University Press, 1991).
Brooke, John Hedley, 'Science and Theology in the Enlightenment', in W. Mark Richardson and Wesley J. Wildman (eds), *Religion and Science: History, Method, Dialogue* (New York and London: Routledge, 1996), 7–27.
Brooke, John Hedley and Ihsanoglu, Ekmeleddin (eds), *Religious Values and the Rise of Science in Europe* (Istanbul: IRCICA, 2005).
Brooke, John Hedley, Osler, Margaret and van der Meer, Jitse, *Science in Theistic Contexts: Cognitive Dimensions*, Osiris, Second Series 16 (Chicago: University of Chicago Press, 2001).
Buck, August, 'Der Italienische Humanismus', in Norbert Hammerstein and August Buck (eds), *Handbuch der deutschen Bildungsgeschichte*, vol. 1: *15. bis 17. Jahrhundert: von der Renaissance und der Reformation bis zum Ende der Glaubenskämpfe* (Munich: Beck, 1996), 1–56.
Büsser, Fritz, 'Melanchthon's *De Providentia*: Entstehung und Aufbau', CR 93/3 (1983), 1–62.
Caspar, Max, *Kepler*, 2nd edn (New York: Dover Publications, 1993).
Cooper, John M., 'Eudaimonism, the Appeal to Nature, and "Moral Duty" in Stoicism', in Stephen Engstrom and Jennifer Whiting (eds), *Aristotle, Kant and the Stoics: Rethinking Happiness and Duty* (Cambridge: Cambridge University Press, 1996), 261–84.
Copleston, Frederick C., *A History of Medieval Philosophy* (Notre Dame and London: University of Notre Dame Press, 1972).
Coyne, George, Hoskin, Michael and Pedersen, Olaf (eds), *Gregorian Reform of the Calendar: Proceedings of the Vatican Conference to Commemorate its 400th Anniversary 1582–1982* (Rome: Pontificia Academia Scientarum, 1983).
Cunningham, Andrew, 'How the *Principia* Got its Name; or, Taking Natural Philosophy Seriously', *History of Science* 29 (1991), 377–92.
Cunningham, Andrew, 'Protestant Anatomy,' in Helm and Winkelmann (eds), *Religious Confessions and the Sciences in the Sixteenth Century*, 44–50.
Cunningham, Andrew and French, Roger, *Before Science: The Invention of the Friars' Natural Philosophy* (Aldershot: Scolar Press, 1996).
Dear, Peter, *Discipline and Experience: The Mathematical Way in the Scientific Revolution* (Chicago and London: University of Chicago Press, 1995).
Dieter, Thomas, *Der junge Luther und Aristoteles: Eine historisch-systematische Untersuchung zum Verhältnis von Theologie und Philosophie* (Berlin: de Gruyter, 2001).
Dilthey, Wilhelm, 'Das natürliche System der Geisteswissenschaften im 17. Jahrhundert', in idem, *Weltanschauung und Analyse des Menschen seit Renaissance und Reformation*, 5th edn, *Gesammelte Werke*, vol. 2 (Göttingen: Vandenhoeck and Ruprecht, 1957), 178–86.
Dingel, Irene, *Concordia controversa: die öffentlichen Diskussionen um das lutherische Konkordienwerk am Ende des 16. Jahrhunderts* (Gütersloh: Gütersloher Verlagshaus, 1996).

Fellermeier, Jakob, 'Das Naturrecht in der Scholastik', *Theologie und Glaube* 58 (1968), 333–69.
Field, J. V., 'A Lutheran Astrologer: Johannes Kepler', *Archive for History of Exact Sciences* 31 (1984), 189–272.
Force, James E. and Popkin, Richard H. (eds), *The Books of Nature and Scripture: Recent Essays on Natural Philosophy, Theology, and Biblical Criticism in the Netherlands of Spinoza's Time and the British Isles of Newton's Time*, International Archives of the History of Ideas 139 (Dordrecht and Boston: Kluwer, 1994).
Forschner, Maximilian, *Die stoische Ethik* (Stuttgart: Klett-Cotha, 1981).
Forstman, H. Jackson, *Word and Spirit: Calvin's Doctrine of Biblical Authority* (Stanford, Calif.: Stanford University Press, 1962).
Frank, Günter, *Die theologische Philosophie Philipp Melanchthons (1497–1560)*, Erfurter theologische Studien 67 (Leipzig: Benno, 1995).
Frank, Günter, 'Gott und Natur: Zur Transformation der Naturphilosophie in Melanchthons humanistischer Philosophie', in Rhein and Frank (eds), *Melanchthon und die Naturwissenschaften seiner Zeit*, 43–58.
Frank, Günter, 'Melanchthon and the Tradition of Neoplatonism', in Helm and Winkelmann (eds), *Religious Confessions and the Sciences in the Sixteenth Century*, 3–18.
Funkenstein, Amos, *Theology and the Scientific Imagination from the Middle Ages to the Seventeenth Century* (Princeton: Princeton University Press, 1986).
García-Ballester, Luis, 'The Inquisition and Minority Medical Practitioners in Counter-Reformation Spain', in Grell and Cunningham (eds), *Medicine and the Reformation*, 156–91.
Geyer, Hans-Georg, 'Welt und Mensch: Zur Frage des Aristotelismus bei Melanchthon', unpubl. doctoral thesis, University of Bonn (1959).
Gilson, Étienne, *History of Christian Philosophy in the Middle Ages* (London: Sheed and Ward, 1955).
Granada, Miguel Angel, 'Il Problema Astronomico-cosmologico e le sacre scritture dopo Copernico: Christoph Rothmann e la "Teoria dell'accomodazione" ', *Rivista di storia della filosofia* 4 (1996), 789–828.
Granada, Miguel Angel, 'Michael Maestlin and the New Star of 1572', *Journal for the History of Astronomy* 38 (2007), 99–124.
Grell, Ole Peter and Cunningham, Andrew (eds), *Medicine and the Reformation* (London and New York: Routledge, 1993).
Greyerz, Kaspar von, Jakubowski-Tiessen, Manfred, Kaufmann, Thomas and Lehmann, Hartmut (eds), *Interkonfessionalität – Transkonfessionalität – binnenkonfessionelle Pluralität: Neue Forschungen zur Konfessionalisierungsthese*, Schriften des Vereins für Reformationsgeschichte 201 (Gütersloh: Gütersloher Verlagshaus, 2003).
Günther, Siegmund, *Peter und Philipp Apian: Zwei deutsche Mathematiker u. Kartographe. Ein Beitrag zur Gelehrten-Geschichte des 16. Jahrhunderts* (facsimile reprod. Amsterdam: Meridian, 1967; 1st publ. Prague: Verlag der königlichen böhmischen Gesellschaft der Wissenschaften, 1882).

Harris, Steven J., 'Confession-Building, Long-Distance Networks, and the Organization of Jesuit Science', *Early Science and Medicine* 1 (1996), 287–318.

Harrison, Peter, *The Bible, Protestantism, and the Rise of Natural Science* (Cambridge: Cambridge University Press, 1998).

Harrison, Peter, 'Newtonian Science, Miracles, and the Laws of Nature', *Journal of the History of Ideas* 56 (1995), 531–53.

Harrison, Peter, 'Voluntarism and Early Modern Science', *History of Science* 40 (2002), 63–89.

Hellman, C. Doris, *The Comet of 1577: Its Place in the History of Astronomy* (New York: Columbia University Press, 1944).

Hellman, C. Doris, 'A Poem on the Occasion of the Nova of 1572', in Edward P. Mahoney (ed.), *Philosophy and Humanism: Renaissance Essays in Honour of Paul Oskar Kristeller* (Leiden: Brill, 1976), 306–9.

Hellyer, Marcus, ' "Because the Authority of My Superiors Commands": Censorship, Physics and the German Jesuits', *Early Science and Medicine* 1 (1996), 319–54.

Hellyer, Marcus, 'Jesuit Physics in Eighteenth-Century Germany: Some Important Continuities', in John W. O'Malley SJ, Gauvin Alexander Bailey, Steven J. Harris and T. Frank Kennedy SJ (eds), *The Jesuits: Cultures, Sciences and the Arts 1540–1773* (Toronto: University of Toronto Press, 1999), 538–54.

Helm, Jürgen, 'Religion and Medicine: Anatomical Education at Tübingen and Ingolstadt', in Helm and Winkelmann (eds), *Religious Confessions and the Sciences in the Sixteenth Century*, 51–68.

Helm, Jürgen and Winkelmann, Annette (eds), *Religious Confessions and the Sciences in the Sixteenth Century* (Leiden: Brill, 2001).

Henry, John, 'The Scientific Revolution in England', in Porter and Teich (eds), *Scientific Revolution*, 115–209.

Hessayon, Ariel and Keene, Nicholas (eds), *Scripture and Scholarship in Early Modern England* (Aldershot: Ashgate, 2006).

Hofmann, Norbert, *Die Artistenfakultät an der Universität Tübingen 1534–1601*, Contubernium 28 (Tübingen: Mohr Siebeck, 1982).

Hooykaas, Reijer, *Fact, Faith and Fiction in the Development of Science*, Boston Studies in the Philosophy of Science 205 (Dordrecht and Boston: Kluwer, 1999).

Hooykaas, Reijer, *Religion and the Rise of Modern Science* (Edinburgh and London: Scottish Academic Press, 1972).

Hoskin, Michael, 'The Reception of the Calendar by Other Churches', in Coyne et al. (eds), *Gregorian Reform of the Calendar*, 255–64.

Howell, Kenneth J., *God's Two Books: Copernican Cosmology and Biblical Interpretation in Early Modern Science* (Notre Dame, Ind.: University of Notre Dame Press, 2000).

Howell, Kenneth J., 'Styles of Science, Calvinism and the Common Good in the Early Dutch Republic', in Brooke and Ihsanoglu (eds), *Religious Values and the Rise of Science in Europe*, 111–30.

Hübner, Jürgen, *Die Theologie Johannes Keplers zwischen Orthodoxie und Naturwissenschaft* (Tübingen: Mohr Siebeck, 1975).

Iliffe, Rob, 'Friendly Criticism: Richard Simon, John Locke, Isaac Newton and the *Johannine Comma*', in Hessayon and Keene (eds), *Scripture and Scholarship in Early Modern England*, 137–57.

Jardine, Nicholas, 'Keeping Order in the School of Padua: Jacopo Zabarella and Francesco Pomponazzi on the Offices of Philosophy', in Kessler *et al.* (eds), *Method and Order in Renaissance Philosophy of Nature*, 183–209.

Kaltenbrunner, Ferdinand, 'Die Polemik über die gregorianische Kalendarreform', *Sitzungsberichte der historisch-philosophischen Classe der kaiserlichen Akademie zu Wien* 87 (1877), 485–586.

Kaltenbrunner, Ferdinand, 'Die Vorgeschichte der gregorianischen Kalenderreform', *Sitzungsberichte der Akademie der Wissenschaften Wien* 82 (1876), 286–414.

Kessler, Eckhard, Di Liscia, Daniel and Methuen, Charlotte (eds), *Method and Order in Renaissance Philosophy of Nature: The Aristotle Commentary Tradition* (Aldershot: Variorum, 1997).

Kopperi, Kari, 'Luthers Theologische Zielsetzung in den philosophischen Thesen der Heidelberger Disputation', *Nordiskt Forum* 1, *Schriften der Luther Agricola Gesellschaft* 28 (1993), 67–103.

Kraye, Jill, 'Moral Philosophy', in Charles Schmitt, Eckhard Kessler and Quentin Skinner (eds), *The Cambridge History of Renaissance Philosophy* (Cambridge: Cambridge University Press, 1988), 303–86.

Kroll, Richard, Ashcroft, Richard and Zagorin, Perez (eds), *Philosophy, Science and Religion in England 1640–1700* (Cambridge: Cambridge University Press, 1992).

Kuhn, Thomas, *The Copernican Revolution: Planetary Astronomy in the Development of Modern Thought* (Cambridge, Mass.: Harvard University Press, 1957).

Kuhn, Thomas, *The Structure of Scientific Revolutions*, 2nd edn (Chicago: University of Chicago Press, 1970).

Kusukawa, Sachiko, '*Aspectio divinorum operum*: Melanchthon and Astrology for Lutheran medics', in Grell and Cunningham (eds), *Medicine and the Reformation*, 33–56.

Kusukawa, Sachiko, 'Lutheran Uses of Aristotle: A Comparison between Philip Melanchthon and Jacob Schegk', in Blackwell and Kusukawa (eds), *Philosophy in the Sixteenth and Seventeenth Centuries*, 169–88.

Kusukawa, Sachiko, 'The Natural Philosophy of Melanchthon and His Followers', in *Sciences et religions de Copernic à Galilée (1540–1610)* (Rome: École Française, 1999), 443–53.

Kusukawa, Sachiko, *The Transformation of Natural Philosophy: The Case of Philip Melanchthon* (Cambridge: Cambridge University Press, 1995).

Kusukawa, Sachiko, '*Vinculum concordiae*: Lutheran Method by Philip Melanchthon', in Kessler *et al.* (eds), *Method and Order in Renaissance Philosophy of Nature*, 337–54.

Leijenhorst, Cees, 'Place, Space and Matter in Calvinist Physics', *Monist* 84 (2001), 520–41.

Leijenhorst, Cees and Lüthy, Christoph, 'The Erosion of Aristotelianism: Confessional Physics in Early Modern Germany and the Dutch Republic', in Cees Leijenhorst, Christoph Lüthy and Johannes M. M. H. Thijssen (eds), *The Dynamics of Aristotelian Natural Philosophy from Antiquity to the Seventeenth Century* (Leiden: Brill, 2002), 375–411.

Lindberg, David C. and Numbers, Ronald L., *God and Nature: Historical Essays on the Encounter between Christianity and Science* (Berkeley and Los Angeles: University of California Press, 1986).

Lohr, Charles H., 'Latin Aristotelianism and the Seventeenth-Century Calvinist Theory of Scientific Method', in Kessler *et al.* (eds), *Method and Order in Renaissance Philosophy of Nature*, 369–80.

Lohr, Charles H., 'Metaphysics and Natural Philosophy as Sciences: The Catholic and Protestant Views in the Sixteenth and Seventeenth Centuries', in Blackwell and Kusukawa (eds), *Philosophy in the Sixteenth and Seventeenth Centuries*, 280–95.

Lohse, Bernhard, 'Dogma und Bekenntnis in der Reformation: Von Luther bis zum Konkordienbuch', in Carl Andresen and Adolf Martin Ritter (eds), *Handbuch der Dogmen- und Theologiegeschichte*, 2nd edn, vol. 2 (Göttingen: Vandenhoeck and Ruprecht, 1998), 1–164.

Lüthy, Christoph, 'An Aristotelian Watchdog as Avant-Garde Physicist: Julius Caesar Scaliger', *Monist* 84 (2001), 520–41.

Maaser, Wolfgang, *Die schöpferische Kraft des Wortes: Die Bedeutung der Rhetorik für Luthers Schöpfungs- und Ethikverständnis* (Neukirchen-Vluyn: Neukirchener Verlag, 1999).

Maaser, Wolfgang, 'Luther und die Naturwissenschaften: Systematische Aspekte an ausgewählten Beispielen', in Rhein and Frank (eds), *Melanchthon und die Naturwissenschaften seiner Zeit*, 25–41.

MacCulloch, Diarmaid, *Reformation: Europe's House Divided 1490–1700* (London: Allen Lane and Penguin, 2003).

Maurer, Wilhelm, *Der junge Melanchthon zwischen Humanismus und Reformation*, 2 vols (Göttingen: Vandenhoeck and Ruprecht, 1967–9; repr. in one vol. with same pagination 1996).

McGrath, Alister, *The Intellectual Origins of the European Reformation* (Oxford: Blackwell, 1987).

McMullin, Ernan, *Galileo, Man of Science* (New York: Basic Books, 1967; repr. Princeton: Scholar's Bookshelf, 1988).

Merton, Robert K., *Science, Technology and Society in Seventeenth-Century England*, 2nd edn (New York: Howard Fertig, 1970).

Methuen, Charlotte, *Kepler's Tübingen: Stimulus to a Theological Mathematics* (Aldershot: Ashgate, 1998).

Methuen, Charlotte, '*Lex naturae* and *ordo naturae* in the Thought of Philip Melanchthon', *Reformation and Renaissance Review* 3. 1–2 (December 2000), 110–25.

Methuen, Charlotte, 'Maestlin's Teaching of Copernicus: The Evidence of his University Textbook and Disputations', *Isis* 87 (1996), 230–47.

Methuen, Charlotte, 'The Role of the Heavens in the Thought of Philip Melanchthon', *Journal for the History of Ideas* 57 (1996), 385–403.
Methuen, Charlotte, 'Securing the Reformation through Education: The Duke's Scholarship System of Sixteenth-Century Württemberg', *Sixteenth Century Journal* 25 (1994), 841–51.
Methuen, Charlotte, 'Special Providence and Sixteenth-Century Astronomical Observation: Some Preliminary Reflections', *Early Science and Medicine* 4 (1999), 99–113.
Methuen, Charlotte, 'Zur Bedeutung der Mathematik für die Theologie Philip Melanchthons', in Reich and Frank (eds), *Melanchthon und die Naturwissenschaften seiner Zeit*, 85–103.
Meyer, Eugen, 'Chronologie IV: Christliche Zeitrechnung' in *Die Religion in Geschichte und Gegenwart*, 3rd edn (Tübingen: Mohr Siebeck, 1957–65), 1, 1814–18
Mikkeli, Heikki, *An Aristotelian Response to Renaissance Humanism: Jacopo Zabarella on the Nature of Arts and Sciences*, Studia Historica 41 (Helsinki: Suomen Historiallinen Seura, 1992).
Moran, Bruce T., 'Princes, Machines, and the Valuation of Precision in the 16th Century', *Sudhoffs Archiv* 61 (1977), 209–28.
Moran, Bruce T., 'Wilhelm IV of Hesse-Kassel: Informal Communication and the Aristocratic Context of Discovery', in Thomas Nickles (ed.), *Scientific Discovery: Case Studies* (Dordrecht: Reidel, 1980), 67–96.
Moran, Bruce T. (ed.), *Patronage and Institutions: Science, Technology, and Medicine at the European Court 1500–1700* (Woodbridge, Suffolk and Wolfeboro, NY: Boydell, 1991).
Moyer, Gordon, 'Aloisius Lilius and the "Compendium rationis restituendi Kalendarium" ', in Coyne *et al.* (eds), *Gregorian Reform of the Calendar*, 171–88.
Nauer, Charles, *Humanism and the Culture of Renaissance Europe* (Cambridge: Cambridge University Press, 1995).
North, John D., *The Universal Frame: Historical Essays in Astronomy, Natural Philosophy and Scientific Method* (London: Hambledon, 1989).
North, John D., 'The Western Calendar – "intolerabilis, horribilis, et derisibilis": Four Centuries of Discontent', in Coyne *et al.* (eds), *Gregorian Reform of the Calendar*, 75–113.
Nutton, Vivian, 'Wittenberg Anatomy', in Grell and Cunningham (eds), *Medicine and the Reformation*, 11–32.
Ohly, Friedrich, 'Deus Geometra: Skizzen zur Geschichte einer Vorstellung von Gott', in Norbert Kamp and Joachim Wollasch (eds), *Tradition als historische Kraft: Interdisziplinäre Forschungen zur Geschichte des früheren Mittelalters* (Berlin: de Gruyter, 1982), 1–42.
Olsson, Herbert, *Schöpfung, Vernunft und Gesetz in Luthers Theologie*, Acta Universitatis Upsaliensis (Studia Doctrinae Christianae Upsaliensia) 10 (Uppsala: University of Uppsala, 1971).
Poole, William, 'Francis Lodwick's Creation: Theology and Natural Philosophy in the Early Royal Society', *Journal for the History of Ideas* 66 (2005), 245–63.

Poole, William, 'The Genesis Narrative in the Circle of Robert Hooke and Francis Lodwick', in Hessayon and Keene (eds), *Scripture and Scholarship in Early Modern England*, 41–57.

Porter, Roy and Teich, Mikulás (eds), *The Scientific Revolution in National Context* (Cambridge: Cambridge University Press, 1992).

Pörtner, Regina, *The Counter-Reformation in Central Europe: Styria, 1580–1630* (Oxford: Clarendon Press, 2001).

Pozzo, Riccardo, 'Die Etablierung des naturwissenschaftlichen Unterrichts unter dem Einfluß Melanchthons', in Rhein and Frank (eds), *Melanchthon und die Naturwissenschaften seiner Zeit*, 273–87.

Rackham, H., 'Introduction', in Cicero, *De Natura Deorum – Academica*, Loeb Classical Library 268 (London and Cambridge, Mass.: Heinemann and Harvard University Press, 1979), pp. vii–xix.

Rex, Richard, 'Humanism', in Andrew Pettegree (ed.), *The Reformation World* (London: Routledge, 2000), 51–70.

Rhein, Stefan and Frank, Günter (eds), *Melanchthon und die Naturwissenschaften seiner Zeit*, Melanchthon-Schriften der Stadt Bretten 4 (Sigmaringen: Thorbecke, 1998).

Rublack, Hans-Christoph (ed.), *Die lutherische Konfessionalisierung in Deutschland: Wissenschaftliches Symposium des Vereins für Reformationsgeschichte 1988*, Schriften des Vereins für Reformationsgeschichte 197 (Gütersloh: Gerd Mohn, 1992).

Schmidt, Heinrich Richard, *Konfessionalisierung im 16. Jahrhundert*, Enzyklopädie Deutscher Geschichte 12 (Munich: Oldenbourg, 1992).

Schöner, Christoph, *Mathematik und Astronomie an der Universität Ingolstadt im 15. und 16. Jahrhundert*, Ludovico Maximilianea Universität Ingolstadt–Landshut–München Forschungen und Quellen 13 (Berlin: Duncker und Humblot, 1994).

Schreiner, Susan, *The Theater of His Glory: Nature and Natural Order in the Thought of John Calvin*, Studies in Historical Theology 3 (Durham, NC: Labyrinth Press, 1991).

Scribner, Robert W., 'Cosmic Order and Daily Life: Sacred and Secular in Pre-Industrial Germany', in Scribner, *Popular Culture and Popular Movements in Reformation Germany*, 1–16.

Scribner, Robert W., *Popular Culture and Popular Movements in Reformation Germany* (London: Hambledon Press, 1987).

Scribner, Robert W., 'Ritual and Popular Religion in Catholic Germany at the Time of the Reformation', in Scribner, *Popular Culture and Popular Movements in Reformation Germany*, 17–47.

Seifert, Arno, 'Das Höhere Schulwesen: Universitäten und Gymnasien', in Notker Hammerstein and August Buck (eds), *Handbuch der deutschen Bildungsgeschichte: 15.–17. Jahrhundert*, vol. 1: *Von der Reformation bis zum Ende der Glaubenskämpfe* (Munich: Beck, 1996), 197–374.

Steinmetz, David, *Calvin in Context* (Oxford and New York: Oxford University Press, 1995).

Stephens, Peter, *The Theology of Huldrych Zwingli* (Oxford: Clarendon Press, 1986).
Stieve, Felix, *Der Kalenderstreit des sechzehnten Jahrhunderts in Deutschland*, Abhandlungen der königlichen bayerischen Akademie der Wissenschaften in München, III. Klasse, Band 15, Abteilung III (Munich: Verlag der königlichen Akademie, 1880).
Strohm, Christoph, 'Beobachtungen zur Melanchthon-Rezeption im frühen Calvinismus', in Johanna Loehr (ed.), *Dona Melanchthoniana: Festgabe für Heinz Scheible zum 70. Geburtstag* (Stuttgart-Bad Cannstatt: frommann-holzboog, 2001), 433–55.
Strohm, Christoph, *Ethik im frühen Calvinismus: Humanistische Einflüsse, philosophische, juristische und theologische Argumentationen sowie mentalitätsgeschichtliche Aspekte am Beispiel des Calvin-Schülers Lambertus Danaeus* (Berlin: de Gruyter, 1996).
Strohm, Christoph, 'Zugänge zum Naturrecht bei Melanchthon', in Günter Frank (ed.), *Der Theologe Melanchthon*, Melanchthonschriften der Stadt Bretten 5 (Stuttgart: Thorbecke, 2000), 339–56.
Taub, Liba Chaia, *Ptolemy's Universe: The Natural Philosophical and Ethical Foundations of Ptolemy's Astronomy* (Chicago and La Salle, Ill.: Open Court, 1993).
Trepp, Ann-Charlott, ' "Nature" and Religious Practice in Seventeenth-Century Germany', in Brooke and Ihsanoglu (eds), *Religious Values and the Rise of Science in Europe*, 81–110.
Wallmann, Johannes, 'Lutherische Konfessionalisierung – Ein Überblick', in Rublack (ed.), *Die Lutherische Konfessionalisierung in Deutschland*, 33–53.
Webster, Charles, 'Puritanism, Separatism, and Science', in Lindberg and Numbers, *God and Nature*, 192–217.
Westman, Robert S., 'The Astronomer's Role in the Sixteenth Century: A Preliminary Study', *History of Science* 18 (1980), 105–47.
Westman, Robert S., 'Humanism and Scientific Roles in the Sixteenth Century', in Rudolf Schmitz and Fritz Krafft (eds), *Humanismus und Naturwissenschaften* (Boppard: Harald Boldt, 1980), 83–100.
Westman, Robert S., 'Johannes Kepler's Adoption of the Copernican Hypothesis', unpubl. doctoral dissertation, University of Michigan, Ann Arbor, 1971.
Westman, Robert S., 'The Melanchthon Circle, Rheticus, and the Wittenberg Interpretation of the Copernican Theory', *Isis* 66 (1975), 164–93.
Whitehead, Alfred North, *Science and the Modern World* (Cambridge: Cambridge University Press, 1926).
Wolf, Ernst, 'Zur Frage des Naturrechts bei Thomas von Aquin und bei Luther', in idem, *Peregrinatio: Studien zur reformatorischen Theologie und zum Kirchenproblem*, vol. 1, 2nd edn (Munich: C. Kaiser. 1962), 183–213.
Wollgast, Siegfried, *Philosophie in Deutschland zwischen Reformation und Aufklärung 1550–1650* (Berlin: Akademie Verlag, 1988).
Wrightsman, Bruce, 'The Legitimation of Scientific Belief: Theory Justification by Copernicus', in Thomas Nickles (ed.), *Scientific Discovery: Case Studies* (Dordrecht: Reidel, 1980), 51–66.

Index

accommodation, doctrine of 2–3, 79, 85–90, 91, 92–3
adiaphora 68–9, 73
Alcuin (735–804) 63
Allen, Diogenes 8, 13
anatomy 20, 26, 95–6, 99
Andreae, Jakob (1528–90) 46n. 45, 51, 52, 54, 105n. 55
Apian, Peter (1495–1552) 41
Apian, Philip (1531–1589) 33, 35, 37, 38, 40–2, 44–6, 56–7, 59, 78
Ariew, Roger 45
Aristarchus of Samos (ca. 310–230 BCE) 83
Aristotle 83, 97, 100–2, 105
  cosmology / cometary theory 38, 41, 46, 55, 56, 83
  De Anima 99, 100
  ethics / *Nichomachean Ethics* 10, 14, 100
  Metaphysica 83, 100
  natural philosophy / *Physica* 13, 38, 95, 100–1
astrology 14n. 30, 40, 51, 54, 92, 98
astronomy 10, 11, 17–21, 27–8, 34–5, 40, 46, 53, 58, 72–3, 77–93, 95, 98–100
Augustine (354–430 CE) 30, 85–6

Bacon, Roger (ca. 1214–94) 63n. 14
Barker, Peter 34, 99n. 22, 108
Bede (673–735) 63
Bernhardt, Reinhold 103–5
Beza, Theodore (1519–1605) 101n. 36, 104
Bible *see* Scripture
Bouwsma, William 8, 13, 17

Brahe, Tycho (1546–1601) 33–4, 82–4, 91, 96n. 8
Brooke, John 49

Cajetan, Thomas de Vio (1469–1534) 105
calendar
  Gregorian reform 61–73
  history of reform 62–5
  Julian 62–3, 68–72
Calvin, John (1509–1564) 2, 8, 17–18, 79, 85n. 30, 92, 100
  and accommodation 89–90
  and Eucharist 107
  and natural world 16–17
  and providence 104–5
Calvinism 18, 77–8, 79, 94–5, 101n. 36, 102, 104, 105, 109, 111
Campanus, Johannes (1220–96) 63n. 14
Christoph, Duke of Württemberg (1515–1568; Duke of Württemberg from 1550) 36n. 11
Cicero (c. 106–43 BCE) 13, 22, 23n. 17, 98
Clavius, Christopher (1538–1612) 64–5, 73
Clement VI (Pierre Roger; 1291–1352; Pope at Avignon from 1342) 64
comets 33–47, 48n. 4, 53–4, 55–8, 73n. 75, 102
  comet of 1577/8 38n. 23, 52n. 16, 53, 58, 62
  comet of 1580 58
  as sublunar or supralunar 41–2, 44
Confessio Augustana (1530) 68
Confessio Augustana variata (1540) 107n. 67
Confessio Belgica (1561) 103n. 42
confessionalization 77–8, 79, 92, 94–5

Copernican hypothesis 83–4, 86, 91, 94–5
Copernicus, Nicolaus (1473–1543) 40n. 29, 57, 83, 86, 88, 94, 98–9, 101–2
Councils
  Fifth Lateran (1512–17) 64
  Fourth Lateran (1215) 106
  of Basle (1431–49) 64
  of Constance (1414–18) 64
  of Nicaea (325) 62, 64n. 19, 65
  of Trent (1545–1563) 64, 67, 106, 109n. 75
creation, doctrine of 1, 8–9, 13n. 21, 14, 16, 28, 30, 49, 52, 55, 58, 73, 78–9, 84, 85–6, 103n. 42, 105
of human beings *imago Dei* 9, 12, 16, 21, 24–5, 26, 53
Cunningham, Andrew 96

Danaeus, Lambertus (1530–1595) 8, 18n. 39, 105
da Vinci, Leonardo (1452–1519) 101
Dee, John (1625–1608/9) 33
Dilthey, Wilhelm (1833–1911) 27
Dürer, Albrecht (1471–1528) 101

Easter
  dating 62–3, 65–6, 68
Eisenmenger, Samuel (also known as Sideocrates; 1534–1585) 35, 38–9, 45
ethics 7–8, 10, 14, 19–20, 24, 105
  see also moral philosophy
Ethics, Nicomachean *see* Aristotle
Eucharist 11
  Calvin and 109
  confessionalization and 95, 106–7
  Kepler and 78
  Luther and 106–9
  Melancthon and 108–9
  transubstantiation 11n. 14, 106–9
  ubiquity 106–9
exegesis *see* Scripture

Fall, results of 9, 11, 12, 16, 17, 22n. 15, 24–5, 27, 30, 54, 58, 89–90, 106
  see also sin
Formula concordiae (1577) 68n. 41, 78, 95
Frischlin, Nicodemus (1547–90) 36n. 11–12, 46
Funkenstein, Amos 107–8

Galen (ca. 129-ca. 216 CE) 13, 95

Galileo Galilei (1564–1642) 96n. 8, 101–2
geometry 21, 25, 80
Gregory XIII: (Ugo Buoncompagno; 1502–1585; Pope 1572–85) 61, 62, 64, 67
  and calendar reform 66–7, 72
Grosseteste, Robert (ca. 1175–1253) 63n. 14

Hafenreffer, Matthias (1561–1619) 51–2, 54
Harrison, Peter 96–8, 99, 101
Heerbrand, Jakob (1521–1600) 55n. 24, 62, 71, 72
  and astrology 54
  and comet of 1577/8 38n. 23, 46n. 48, 52–3
  and Gregorian calendar 67–9, 73
  and *liber naturae* 52–3, 97n. 13
  and providence 51, 52–4
  calendar as adiaphora 68–9, 73
Hübner, Jürgen 77–8, 79
hypothesis 2–3, 79, 82–3, 91, 92
imago Dei see creation, doctrine of

Inquisition 85

James VI and I (1566–1625; King of Scotland from 1567; King of England from 1603) 80, 81
Javelli, Chrisostomo (1470–1538) 105
Jesuits 64, 102, 105, 109
John XXIII (Baldassare Cossa; ca. 1370–1419; schismatic Pope 1410–15) 64

Kepler, Johannes (1571–1630) 2, 33, 52, 97n. 10, 102
  and accommodation 89–91
  Astronomia Nova 82–8
  and confessionalization 77–8, 79, 81
  and Copernicus 83–4, 86
  and Eucharist 78
  Harmonices Mundi 80–1, 88
  and hypotheses 84, 91
  and *liber naturae* 79–80, 81, 85–90
  as Lutheran astronomer 92–3
  Mysterium Cosmographicum 79, 80, 81–2, 87–9
  as Reformer of astronomy 79, 81–3
  and Scripture 85–90

Index

Kusukawa, Sachiko 2, 95–6, 99, 102–4

Lactantius (ca. 250-ca. 325 CE) 85–6
Last Days / Final Judgement 39, 43, 56, 66, 71
Leo X (Giovanni di Lorenzo de' Medici; 1475–1521; Pope from 1513) 64
Leovitius, Cyprian (1524–74) 33, 35, 36–7, 38–9, 40
Liebler, Georg (1524–1600) 100–1
Lilius of Verona, Aloysius (1510–76) 64
Lohr, Charles 104n. 52, 105–6
Ludwig, Duke of Württemberg (1554–93; Duke from 1568) 34–6, 56–7
  correspondence with Wilhelm of Hesse 34–46
Luther, Martin (1483–1546) 2, 7, 13, 17–18, 20, 30n. 57, 51, 65–6, 67, 68–9, 71–2, 73, 79, 88, 92–3, 96, 110
  and the calendar 65–7
  and the Eucharist 11n. 14, 107–9
  Law and Gospel 17
  and the limits of reason 10–11, 88n. 42
  and the natural world 9–10, 100
  and the order of society 10–11
  and philosophy 10–11, 13, 18, 100
  and providence 103–5

Maestlin, Michael (1550–1631) 36n. 12, 55n. 24, 56–9, 62, 78n. 6, 80n. 11, 87, 92, 96n.8
  and 1572 nova 33, 35, 40, 44–6, 56–7
  and comets 58
  and Gregorian calendar 69–73
mathematics / mathematical sciences 12–3, 19–23, 23–9, 34, 43, 53, 58, 71–3, 80, 81, 84, 90, 91, 98,
Maximilian I (1459–1519; Holy Roman Emperor from 1508) 64
Melanchthon, Philip (1497–1560) 2, 7, 15, 16, 40, 52, 79, 80, 92–3, 94
  and anatomy 96
  and astrology 51, 54
  and Copernicus 98–99, 102
  and Eucharist 107–9
  Initia doctrinae physicae 12, 26, 53, 98, 99n. 22, 100–1
  Loci communes 12–3, 21–3, 23–4, 25, 27, 28

Lutheran natural philosophy 2, 95–6, 98–101, 110–11
  and natural law 12, 21–3, 25–9
  and order of nature 11–13, 21, 27–9, 53
  and philosophy 20–1, 23–9
  proofs of existence of God 12–13, 19n. 2, 28
  and providence 53–4, 102–6
  and reason 13, 17–18
  and the teaching of philosophy 100
Merton, Robert 94
moral philosophy 9, 10–11, 12–15, 17–18, 19, 23–6, 28–9 81, 92, 100, 108
  relationship to natural philosophy *see* natural philosophy

natural law 2, 9, 10, 12–13, 15, 19–30
natural philosophy 2, 7, 9, 10–11, 12–13, 17–18, 19, 20, 25–9, 34, 36, 44–6, 85, 86, 88, 92, 94–111
  Aristotelian 36, 44–6, 56
  Lutheran 2–3, 94, 95–6, 94–111
  relationship to moral philosophy 9–10, 14–15, 17–18, 23–9, 92–3, 98n. 17, 110
natural order 8, 10, 11, 13, 14, 17, 28, 54
natural theology 9, 10, 105, 106n. 60
Nicolas of Cusa (1401–64) 64
nova (1572) 2, 33–47, 49, 55–9, 102
  distance from earth 40–3, 44, 55
Nutton, Vivian 99

Old Testament
  Genesis 9, 86, 90
  Psalms 9, 86
  wisdom tradition 9, 48, 58
order of nature/creation 2, 8–9, 9–11, 12–13, 16, 26, 28, 48, 50–5, 58, 99

Patrizzi, Francesco (1529–97) 102
Paul of Middelburg (1446–1534) 64
Peace of Augsburg (1555) 95
Perera, Benedict (also Pereira or Pereyra; 1535–1610) 105
Peucer, Caspar (1525–1602) 19, 33, 35, 37, 44–6, 56
Pico della Mirandola, Giovanni (1463–1494) 51

Pius IV (Giovanni Angelo Medici; 1499–1565; Pope from 1559) 64
Plato (ca 423-ca 347 BCE) 13, 21, 22, 25, 28, 30, 53, 80, 81, 84, 101
planets
  Jupiter 38, 39, 40, 41, 57
  Mars 38, 39, 82n. 21, 83
  Mercury 98
  Saturn 40, 41, 57, 90
  Venus 39, 42–3, 44, 56n. 24, 98
Pliny (ca. 23–79 CE) 14, 41
Proclus (ca. 412–85 CE) 80
providence, doctrine of 1–2, 8–9, 13, 14–15, 16, 17, 43–4, 48–60, 89, 102–6
  general (*providentia generalis*) 15n. 38, 43, 48–54, 58–60, 103, 105–6
  special (*providentia specialis*) 15, 17, 43, 44, 48–60, 105
Ptolemy (ca 85–160 CE) / Ptolemaic worldview 2, 13, 41, 55, 56, 82, 84, 91, 95, 98–9
Puritans/Puritanism 94
Pythagoras (5th century BCE) 53, 84

Regiomontanus, Johannes (1436–76) 64
Romans, Epistle to 9, 13n. 21, 22, 89
Rothmann, Christoph (ca. 1560-ca. 1600) 105

Sacro Bosco, John (†1244 or 1256) 63n. 14
Scaliger, Joseph Justus (1540–1609) 66n. 30, 73
Schegk, Jacob (1511–1587) 101n. 35
Schreiner, Susan 8,
Scripture 3, 20, 22, 23, 24, 61–2, 71, 79, 81, 85–91, 92, 96, 104, 105, 110
  and law 21–4
  liber scripturae and *liber naturae* 48, 52, 85–9, 93, 97n. 13
  reading of 20, 86, 96n. 9,
Sideocrates *see* Eisenmenger, Samuel
sin 7, 8, 9, 11
Sixtus IV (Francesco della Rovere; 1414–1484; Pope from 1471) 64
Stieve, Ferdinand 67

Stoics / Stoic 17, 30, 92, 98, 107, 108
Strohm, Christoph 8, 13, 21, 104
Suarez, Francisco (1548–1617) 105
sublunar world 11, 21n. 8, 27, 42, 44
supernova (1572) *see* nova (1572)

Telesio, Bernardino (1509–88) 102
Thales of Miletus (ca. 625-ca. 546 BCE) 98
Thirty Years' War 78, 80
Thomas Aquinas (1225–74) 30, 50, 106

Ulrich, Duke of Württemberg (1487–1550; Duke of Württemberg from 1495) 36n. 11
universities:
  Frankfurt-an-der-Oder 100
  Greifswald 100
  Heidelberg 62, 100
  Ingolstadt 36n. 12
  Königsberg 100
  Leuven 64
  Marburg 100
  Padua 97n. 32
  Rostock 100
  teaching of moral and natural philosophy at 18, 100–1
  Tübingen 2, 18n. 48, 33, 35–6, 42n. 36, 51, 52, 62, 64, 97n. 13, 100
  Vienna 64
  Wittenberg 33, 56, 94, 95

Vesalius, Andreas (1514–64) 96, 101n. 35

Westman, Robert 95
Wilhelm IV, Landgrave of Hesse (1532–92; Landgrave from 1567) 33–4, 35, 37, 42–3, 43–6, 56
  correspondence with Ludwig of Württemberg 34–46
Württemberg, Duchy of 2, 33, 34
  assent to *formula concordiae* 78

Zoestius, Hermann (1380–1445) 64
Zwingli, Huldrych (1484–1531) 2, 8, 14–15, 16, 17–18, 78, 103–5, 107–8